THE OUTCOMES AND APPLICATIONS OF HAI RIVER BASIN INTEGRATED WATER AND ENVIRONMENT MANAGEMENT PROJECT

Foreign Economic Cooperation Office, Ministry of Environmental Protection

China Irrigation and Drainage Development Center

Hai Water Conservancy Commission, Ministry of Water Resource

China Environment Press

图书在版编目（CIP）数据

海河水资源与水环境综合管理项目研究成果与应用＝The Outcomes and Applications of Hai River Basin Integrated Water and Environment Management Project：英文 / 环保部环境保护对外合作中心，中国灌溉排水发展中心，水利部海河水利委员会编著. — 北京：中国环境出版社，2012.10
 ISBN 978-7-5111-0617-9

Ⅰ.①海… Ⅱ.①环…②中…③水… Ⅲ.①海河－流域－水资源管理－研究－英文②海河－流域－水环境－综合管理－研究－英文 Ⅳ.① TV213.4 ② X143

中国版本图书馆 CIP 数据核字（2011）第 116901 号

策划编辑	王素娟
责任编辑	俞光旭
责任校对	唐丽虹
封面设计	陈 莹
排版制作	杨曙荣

出版发行	中国环境出版社
	（100062 北京市东城区广渠门内大街16号）
网　　址	http://www.cesp.com.cn
联系电话	010-67112765（编辑管理部）
发行热线	010-67125803　010-67112705
印装质量热线：010-67113404	
印　　刷	北京盛通印刷股份有限公司
经　　销	各地新华书店
版　　次	2012年12月第1版
印　　次	2012年12月第1次印刷
开　　本	787×960　1/16
印　　张	15
字　　数	250千字
定　　价	86.00元

【版权所有。未经许可，请勿翻印、转载，违者必究】

Full List of Contributors

Chief Editor:

Li Pei   Han Zhenzhong   Tang Yandong   Liu Bin

Zhang Xiaolan   Li Yandong   Yang Yuchuan   Chang Mingqi

Authors (In alphabetical order by last name):

Che Hongjun   Chen Ying   Cheng Yanping   Dong Hansheng   Fu Guo   Fu Jian
Fu Xiaoliang   Gao Fei   Ge Chazhong   Gu Yangwen   Gu Tai   He Hao
He Ping   He Yunya   Huang Ren   Huang Jinlin   Jiang Yunzhong   Li yonggen
Li Chunhui   Li Huichang   Li Jia   Li Jianxin   Li Qianxiang   Li Wei   Li Wei
Lin Chao   Liu Hongxian   Liu Shengbin   Liu Yu   Lu Xueqiang   Luo Zunlan
Ma Jiyuan   Ma Ming   Ma Wei   Meng Haiyang   Meng Xianzhi   Peng Junling
Qi Wenjie   Shen Dajun   Song Qiubo   Su Baolin   Sun Ren   Sun Minzhang
Wang Liming   Wang Liqing   Wang Shuqian   Wang Yue   Wang Yunkai
Wang Zhiliang   Wang Zhongjing   Weng Wenbin   Wei Wenting   Wu Bingfang
Wu Shunze   Wu Xiaopu   Xu Lei   Xu Linbo   Xu Yi   Xu Zongxue   Yan Li
Yan Nana   Yan Xuejun   Yang Qian   Yu Jingshan   Yu Weidong
Yu Xiangyong   Yuan Caifeng   Zang Yuxiang   Zhang Junxia   Zhang Xisan
Zhang Yuan   Zheng Hongqi   Zheng Jun   Zhong Yuxiu   Zhou Zuhao
Zhu Xiaochun   Zhu Xinjun

Techmcal Advisory:

Meng Wei   Ren Guangzhao   Ma Zhong   Su Yibing   Xu Xinyi

Zhang Guoliang   Zhao Jingcheng   Xiao Qing   Jiang Liping

Douglas Olson   Edwin Ongley   Clive Lyle   Wim Bastiaanssen

Peter Droogers   Tim Bondelid   Richard Evans

Foreword

Sustainable development demand in the Haihe River Basin is pressing due to the severe water resource and environmental problems in the basin. Since the 9th Five-Year Plan, the Haihe River Basin has always been kept on the priority of the state, the Ministry of Environmental Protection and the Ministry of Water Resources for pollution prevention, one of water pollution prevention and control basins of "Three Major Rivers and Three Major Lakes". To effectively alleviate water resource shortage, restore the ecological environment and reduce the pollution of basin land sources to the Bohai Sea and really improve the water environment, the Ministry of Environmental Protection and the Ministry of Water Resources began to design the GEF Hai River Basin Water Resources and Water Environment Management Project (hereinafter referred to as the "GEF Hai River Project") in 2002 with the strong support from the Global Environment Facility and the World Bank. The Ministry of Environmental Protection and the Ministry of Water Resources jointly initiated the GEF Hai River Project in September 2004. As for the project fund, 17 million dollars were donated by the GEF and financial departments of various levels invested 17.65 million dollars. Established objectives of the project have been realized under the joint efforts of parties concerned 8 years since the implementation of the project. The project explored the innovative integrated basin management mode in the Hai River Basin, which created an innovative basin integrated management mechanism involved in both domestic and overseas cross-department multi-disciplinary experts and scholars, thus setting a good example for domestic environmental protection department and water resource department to carry out basin integrated management cooperation.

Reviewing the implementation of the GEF Hai River Project, the following contributed a lot to the success of the project: Firstly, high importance the Ministry of Environmental Protection and the Ministry of Water Resources had attached to provide a strong organization guarantee. The Ministry of Environmental Protection and the Ministry of Water Resources established a Project Guidance Committee for the comprehensive management of water resources and water environment of the GEF Hai River Basin. The vice minister in charge was appointed as the director of the committee and leaders of related departments as well as leaders of the Environmental Protection Department (Bureau) of Beijing, Tianjin and Hebei were appointed as committee members. The Committee took charge of and coordinated works of the project, which provided strong organization guarantee for the successful implementation of the project. The Pollution Prevention and Control Department of the Ministry of Environmental Protection, as the Zhangweinan Sub-Basin Project Coordination Leading Group Unit and the Water Resources Department of the Ministry of Water Resources as the Zhangweinan Sub-Basin Project Coordination Vice-Leading Group Unit were involved in the whole implementation of the project. They played a crucial role in the successful implementation of the project of the Zhangweinan Sub-basin until the whole Hai River and laid a solid foundation for the subsequent promotion of the project achievement application. Secondly, the project employed over one hundred famous experts and scholars home and abroad to establish a technical think tank. Domestic experts involved were from Chinese Academy of Sciences, Chinese Research Academy of Environmental Sciences, Chinese Academy for Environmental Planning, Tsinghua University, Peking University, China Agricultural University, China Institute of Water Resources and Hydropower Research and other well-known research institutes and universities; overseas experts involved were from the United States, Canada, Australia and Europe with rich experiences in basin management. They had conducted extensive exchanges and consultations as well as seminars with the domestic project implementation unit focusing on the integrated management plan of water resources and environment, knowledge management, water conservation management, small city waste water management, basin non-point source pollution control, monitor and evaluation, which had raised the project management level and technology content comprehensively. Thirdly, project organization management and project results had innovative and practical application value. The GEF Hai River Project created a precedent for the domestic environmental protection department and water resource department to cooperate and implement projects. Such cross-department cooperation mechanism created and operated in this project set a good example for domestic cross-department cooperation; pushed by the project, the Ministry of Environmental Protection and the Ministry of Water Resources signed a Hai River Basin Data Sharing Agreement, which greatly promoted the integrated management of domestic basin. Integrated basin management based on water consumption management was carried out from the basin level and non-point source pollution prevention and control research were carried out in cooperation with the agricultural department for the first time. The project preparation team completed 8 Integrated Management and Strategy Research on Water Resources and Water Environment. They also completed 2 Integrated Management and Strategic Action Plans on Water Resources and Water Environment for the Haihe river basin and Zhangweinan sub-basin. Additionally, they completed 17 Integrated Management Plans On Water Resources And Water Environment and more than 10 platforms for Knowledge Management System and 6 water-conservation and projects for pollution reduction demonstration. These results had been widely publicized both home and abroad in the form of international thesis released on international conferences and aroused wide attention from domestic and foreign counterparts, and played an active role in the actual work in the environmental protection department and the water resources department of various levels. The above achievements were widely used in the implementation of the 11th Five-Year Plan and the preparation of the 12th Five-Year Plan of the Ministry of Environmental Protection and the Ministry of Water Resources, Beijing, Tianjin and governments of various levels in Hebei Provinces. Fourthly, a batch of water resources and water pollution control and prevention technical experts and project management personnel with international standards had been cultivated through project implementation and the capacity of units involved in project construction had been improved comprehensively, laying a solid foundation for future all-round integrated management on Hai Rvier Basin Water Resources and Water Environment.

The book is the fruit of hard work of hundreds of research and project management personnel involved in the GEF Hai River Project. Only main compilers of the book are listed. Due to the objective conditions, it is difficult to include all experts and representatives involved in the project management and research in the book. Here we would like to take advantage of this book-publishing opportunity to extend our sincere appreciation to leaders and experts from the Global Environment Facility, the World Bank, the Ministry of Finance, the Ministry of Environmental Protection and the Ministry of Water Resources involved in the GEF Hai River Project. Our gratitude also goes to experts, scholars and technical personnel involved in the research work of the project as well as experts and scholars home and abroad who have given support to the project. As internationally advanced experiences and technologies were introduced in the project, studying and application of new technologies and methods can inevitably avoid inadequacy. We hope that experts and scholars can give us more criticism and guidance.

Editor

October 2012

TABLE OF CONTENTS

Chapter 1 OVERVIEW OF HAI RIVER BASIN AND BOHAI SEA 1

1.1 BOHAI SEA — The Context 1
    1.1.1 Physical and Biological Environment of Bohai Sea and Bohai Bay 1
    1.1.2 Exploitation of Natural Resources of Bohai Sea 3
    1.1.3 Environmental Pressures on Bohai Sea and Bohai Bay 4
1.2 HAI RIVER BASIN 8
    1.2.1 Socio-economic Profile.. 9
    1.2.2 Environment 10
    1.2.3 Evapotranspiration (ET).. 16
    1.2.4 Social and Economic Situation in Hai River Basin 18
1.3 INSTITUTIONAL SITUATION AND LEGAL FRAMEWORK 20
    1.3.1 The Basin Management System 20
    1.3.2 Analysis of the Problems of the Present Management System 23
1.4 PLANNING SYSTEM IN HAI RIVER BASIN 28
    1.4.1 Plan Types 28
    1.4.2 Planning Procedures 29

Chapter 2 PROJECT CONTENT AND TECHNOLOGY ROADMAP 32

2.1 THE GEF HAI RIVER PROJECT 32
2.2 PROJECT OBJECTIVES 33
    2.2.1 Geographical Scope... 33
    2.2.2 Project Goals... 34
    2.2.3 Specific Objectives 35
2.3 OVERALL PROJECT TECHNICAL FRAMEWORK 36
2.4 MANAGEMENT OF THE PROJECT 36
2.5 TECHNICAL QUALITY 37
2.6 PEFORMANCE MANAGEMENT 39
    2.6.1 Project Monitoring and Evaluation Management System...................... 39
    2.6.2 Indicator System for Project Monitoring and Evaluation....................... 40

2.7　CAPACITY BUILDING 41
　2.7.1　Top-Down, Bottom-Up and Horizontal Management Principles 41
　2.7.2　Training 41
　2.7.3　Baseline Survey 42

Chapter 3 CONCEPTUAL INNOVATIONS AND THREE TECHNOLOGIES 46

3.1　NEW CONCEPTS -IWEM 46
　3.1.1　IWEM in Action 47
3.2　NEW TECHNOLOGIES IN GEF HAIHE PROJECT 47
　3.2.1　Remote Sensing for ET Management 47
　3.2.2　Knowledge Management (KM) 49
3.3　River Coding 59
　3.3.1　The Stream Reach as the Fundamental Unit of a Surface Water Knowledge Base 59
　3.3.2　Characteristics of the River Coding System 60
　3.4　ET Management 60
　3.4.1　Remote Sensing Estimate of ET 63
3.5　Development and Application of the SWAT Model and Dualistic Model 72
　3.5.1　SWAT Model in Zhangweinan Sub-basin 72
　3.5.2　SWAT Model of Hai River Basin 73
　3.5.3　Development and Application of Dualistic Model 75

Chapter 4 STRATEGIC STUDIES FOR WATER RESOURCES AND WATER ENVIRONMENT AND THE DEMONSTRATION PROJECTS 80

4.1　INTRODUCTION 80
4.2　STRATEGIC STUDY OF INSTITUTIONAL MECHANISMS, POLICIES AND REGULATIONS 81
　4.2.1　Overall Objective 81
　4.2.2　Suggestions for IWEM Policies in Hai River Basin 81
　4.2.3　Improving the Legal Framework 83
　4.2.4　Suggestions on Institutional Reform for IWEM in Hai River Basin 84
　4.2.5　Implementation Plan for Reform 85

4.3　STRATEGIC STUDIES ON WATER RESOURCES 87
  4.3.1　SS4: Efficient Water Use and Water Savings 87
  4.3.2　SS5: Sustainable Utilization of Groundwater, Water rights, and Water Withdrawal Permitting 90
  4.3.3　SS6: Wastewater Recycling 94
  4.3.4　SS8: Reasonable Water Allocation in Beijing After the South–North Water Transfer 97
4.4　STRATEGIC STUDIES ON WATER POLLUTION CONTROL 98
  4.4.1　SS2: IWEM Strategy for the Bohai Sea 98
  4.4.2　Strategic Study 3: Water Ecological Restoration 106
  4.4.3　Strategic Study 7: Control of Pollution Sources 109
  4.4.4　Rural Non-Point Source(NPS) Pollution 115
4.5　DEMONSTRATION PROJECTS 120
  4.5.1　Demonstration Projects of Water Resources Management 120
  4.5.2　Water Pollution Control Demonstration Projects 127

Chapter 5　INTEGRATED WATER RESOURCES AND ENVIRONMENT MANAGEMENT PLANS (IWEMPs) 135

5.1　OVERVIEW OF INTEGRATED WATER RESOURCES AND ENVIRONMENT MANAGEMENT PLANNING 135
5.2　TIANJIN MUNICIPALITY IWEMP 136
  5.2.1　Tianjin Municipality Overview 136
  5.2.2　Highlights of Tianjin Municipality's IWEM Plan 137
5.3　TIANJIN MUNICIPALITY IWEMP SPECIAL STUDIES 143
  5.3.1　Special Study on Water Quality 143
  5.3.2　Water Ecology and Rehabilitation 151
5.4　TYPICAL COUNTYLEVEL IWEMPLANS 159
  5.4.1　Pinggu District of Beijing Municipality 159
  5.4.2　Xinxiang County of Henan Province 169
  5.4.3　Guantao County of Hebei Province 171

Chapter 6 STRATEGIC ACTION PLAN (SAP) AT THE BASIN AND SUB-BASIN LEVEL 179

    6.1   GENERAL INTRODUCTION ABOUT SAP 179
    6.2   SAP AT THE LEVEL OF HAI RIVER BASIN 181
        6.2.1   Basin Features... 181
        6.2.2   Socio-economic Characteristics.. 182
        6.2.3   Main Problems.. 182
        6.2.4   Tasks and Objectives for the SAP at Hai River Basin Level 185
        6.2.5   Principal Recommendations of the Hai River Basin SAP..................... 187
        6.2.6   Safeguard Measures for SAP at Hai River Basin Level 189
    6.3   SAP FOR ZHANGWEINAN SUB-BASIN.. 190
        6.3.1   General Introduction of the Zhangweinan Sub-basin............................ 190
        6.3.2   Zhangweinan Sub-basin SAP 193

Chapter 7   SUMMARY OF PROJECT OUTOMES AND EXPERIENCE 201

    7.1   CONTEXT OF THE PROJECT... 201
    7.2   MONITORING AND EVALUATION... 202
    7.3   PROJECT OBJECTIVES... 202
    7.4   MAIN OUTCOMES OF THE PROJECT 203
    7.5   PROJECT IMPLEMENTATION EXPERIENCE 207
        7.5.1   Cross-departmental Cooperation Mechanism.................................... 208
        7.5.2   Key Innovations... 212
    7.6   THE GEF HAI RIVER PROJECT AND THE 12TH "FIVE- YEAR" PLAN 221

ABBREVIATIONS and ACRONYMS

BOD_5	Biological Oxygen Demand (over a 5 day period)
COD	Chemical Oxygen Demand
CPMO	Central Project Management Office
DIN	Dissolved inorganic nitrogen
DSS	Decision support system
EPB	Environmental Protection Bureau (local levels)
ET	Evapotranspiration
GEF	Global Environment Facility (UN "green fund")
HWCC	Haihe Water Conservancy Commission (Hai river basin commission)
IEP	International Expert Panel
IWEM	Integrated Water and Environment Management
IWRM	Integrated Water Resource Management
CJEG	Central Joint Expert Group
KM	Knowledge Management
m^3	cubic meters
M&E	Monitoring and Evaluation
MEP	Ministry of Environmental Protection
MIS	Management Information System
MWR	Ministry of Water Resources
NH_3	Ammonia
NPS	Non-point Source (pollution)
PAD	Project Appraisal Document
ppm	Parts per million
RS	Remote Sensing
SNWT	South to North Water Transfer Scheme
TDP	Total Dissolved Phosphorus
ToR	Terms of Reference
TUDEP	Tianjin Urban Development & Environment Project
WRPB	Water Resources Protection Bureau (part of HWCC)

Chapter 1

OVERVIEW OF HAI RIVER BASIN AND BOHAI SEA

1.1 BOHAI SEA — The Context

The driving force behind this GEF project is the deteriorated status of Bohai Sea caused mainly by loss of freshwater input to the Sea, high levels of pollution, and consequent ecosystem dysfunction.

1.1.1 Physical and Biological Environment of Bohai Sea and Bohai Bay

Bohai Sea is a partially enclosed inland sea in the northeastern part of China (Figure 1.1) connected to the Yellow Sea. It stretches from Laotie Mountain in Liaodong Peninsula in the northeast to the Shandong Peninsula (Penglai angle) in the south. The sea covers an area from west 117°32'E to east 122° 08'E, and from 37° 07' N to 40 °55' N. This encompasses sea area of 77,000 km² with northeast to southwest distance about 555 km and east-west width of 346 km. Bohai Sea has total coastline length about 3,780 km, which there is approximately 3,020 km of land coastline. It is surrounded by four provinces of Liaoning, Hebei, Tianjin

Figure 1.1 General map of Bohai Sea (A) and surrounding area (from Google maps)

and Shandong. Bohai Sea has an average depth about 18 m with maximum depth about 80 m. Bohai Sea consists of five parts which are Liaodong Bay in the north, Bohai Bay in the west, Laizhou Bay in the south, the central basin in the middle and Bohai Strait in the east that connects Bohai Sea to the Yellow Sea. Few of the rivers entering the gulf include the Yellow River, Hai River, Liao River, and Luan River. Bohai Bay and Liaodong Bay receive the drainage from the Hai River Basin. Details of these areas are found in Table 1.1.

Table 1.1 Marine areas of Bohai Sea

Location	Area/km^2	Average Depth/m	Maximum Depth/m	Coastal Rivers
Liaodong Bay	18,000	22	32	Liao River
Bohai Bay	12,500	20	26	Hai River
Laizhou Bay	7,400	13	—	Coastal River
Central basin	—	20 ~ 25	30	Liao River

The Miao Island Archipelago is oriented north-south in the central and southern parts of the Bohai Strait, dividing the strait into 12 waterways. These waterways vary from each other in width and depth. Generally, these are wider in the north and narrower in the south, deeper in the north and shallower in the south. The deepest and widest waterway in the northern part is the Laotieshan Waterway with a maximum depth of 80 m. The Bohai Sea contains 406 islands with 268 larger than 500 m^2. There are vast areas of shallow sea and tidelands in Bohai Sea, as well as tens of thousands of hectares of coastal areas of low-lying saline-alkali soil. These areas are some of the most actively developed and most valuable resources for marine development. Bohai Bay, with a small mouth, is relatively weak in water exchange. There are many studies and reports about the time needed for full water exchange cycle in Bohai Sea; these range from 16 years, 40 years or 160 years however these is no consensus on this matter.

There are more than 600 species of life in the Bohai Sea, including >120 species of phytoplankton with an annual primary productivity of 112 mg/m^2; >100 kinds of zooplankton, >100 kinds of intertidal benthic plants, >140 kinds of intertidal benthic fauna, >200 kinds of sub-tidal shallow water benthic fauna, and >120 kinds of aquatic animals. There are 5 families and 27 species of fish, as well as shrimp, sea cucumber, abalone and other seafood. The entire Bohai Sea is suitable for fishery. The main farmed species along the Bohai Sea include seaweed, shellfish and shrimp, sea cucumber, abalone and so on .

1.1.2 Exploitation of Natural Resources of Bohai Sea

Marine Traffic: Bohai Sea has many bedrock harbors with deep water and has more than 70 coastal sections suitable for harbor building. There are currently 66 harbors completed or under construction, including 48 fishing ports. It is China's most port-intensive area with the key ports such as Dalian, Qinhuangdao, Tianjin, Yantai, Yingkou, etc. Throughput from Bohai Sea ports accounts for 45% of the country 's major ports.

Oil and Gas Production: There are rich oil resources in the Bohai Bay area which is China's second largest oil-production area. The oil and gas resources are mainly located in coastal, land and offshore continental shelves. The basin covers all or parts of ZhongYuan Oil-field, North China Oil-field, Dagang Oil-field, and Shengli Oil-field. It has about 1.5 billion tons of reserves, with an annual extraction volume of 36 million tons .

Salinity: Bohai Sea has a high salinity of >30‰ ; evaporation is greater than precipitation. There are broad tidal flats and rich brine resources, all of which make it suitable for the salt industry. There are 16 salt fields with 1,600 km^2 of salt works areas. The salt fields that could be developed and utilized are up to 2,500 km^2, which make the Bohai Sea the country's largest salt production base.

Minerals: The area surrounding Bohai Sea is also rich in mineral resources. Reserves of coal, oil, natural gas, iron, aluminum, gypsum, graphite, sea salt, etc. are among the largest in the country. Especially for coal, reserves are thought to be up to 202.6 billion tons, accounting for 45% of the country's total reserves; annual output is 280 million tons accounting for 20% of the national output.

Tourism: The Bohai area is rich in tourism resources and is famous for mountains, ocean sceneries and cultural heritage. Well-known tourist sites include the Great Wall, the Watertown, and the Palatinate for Emperor Qin, the Imperial Palace complex, etc..

Fishery: There has been a decrease in fish species and its diversity in Bohai Sea. The fishery decline in Bohai Sea is mainly caused by over-fishing, reduction of runoff into the sea, rising of salinity and environmental pollution. Reduction of fishery resources is mainly characterized by the substantial reduction of high economic value demersal fish and the decline in its proportion of the total catch. Also important species are decreasing and populations are younger and less-valued.

1.1.3 Environmental Pressures on Bohai Sea and Bohai Bay

1.1.3.1 Introduction

Since reform and opening up, the coastline has been transformed by port construction, large-scale shore aquaculture, dredging, dike construction, land reclamation, oil exploration, etc.. Development has had a major impact on coastal organisms, as well as on habitat and ecosystems. Since the 1970s excessive development of fishing capacity and fishing intensity have exceeded resource renewal capacity and has resulted in continual decrease in fish production. Over-fishing also had a negative impact on the ecological balance of the marine system, resulting in the decrease of the major economic fish species, especially the reduction of the demersal fish resources. Now, the main fish resources in the sea area near Tianjin have been reduced to such an extent that the original four fishing seasons no longer exist. The target of ecological restoration cannot be attained through the current efforts on limiting fishing and natural reproduction.

The flux of large amount of sewage and harmful substances into the sea from Hai River Basin and other basins around Bohai Sea have resulted in severe eutrophication, frequent red tides, ecological imbalance, habitat destruction and reduction in biodiversity and biological resources. Because of the weak water exchange capacity of Bohai Sea and Bohai Bay, both suffer from weak self-purification and assimilative capacities.It is recognized that this situation seriously hampers the sustainable development of the regional economy.

1.1.3.2 Physical Impacts on Bohai Sea

Physical impacts include the decreasing amounts of freshwater and sediment transported to the sea, coastal development, and other human activities such as over-fishing.

Runoff: There has been a sharp reduction in runoff to the sea since the 1960s. The water utilization in Hai River Basin is more than 100 % (due to water transferred from outside the basin and the desalination of sea water). With the reduction in runoff from the rivers into the sea, the transfer of river sediment load is also reduced accordingly resulting in shoreline erosion.Reduction in the freshwater net flux into the sea has resulted in an increase in the average salinity of Bohai Bay. Over the past half a century the average salinity of the entire Bohai Sea has been increased by nearly 2 ‰, and Bohai Bay has seen the largest increase in salinity, with an increase of around 10 ‰ in the Laohuang estuary of Bohai Bay.

Coasts: Under the influence of human activities, the ecological service function of coasts has significantly declined; the natural evolution of lagoons has been changed to varying degrees, and human activities directly determine the direction of lagoon development or demise.

Sediment: The Bay sediments are mainly from river sediment. However, with the decrease of runoff from coastal rivers, sediment loadis also correspondingly reduced. Especially from the Haihe River, where there has been almost no water and sediment flow into the sea since the late of 1970s. The original sediment dynamic balance has been destroyed.

1.1.3.3 Pollution Damage to the Bohai Sea

Water quality problems: Bohai Bay is heavily polluted by nitrogen, phosphorus, and oil pollutants. Severe algal growth along the coast is a major problem (Figure 1.2). The oil contaminated areas are mainly waters near Dagukou Tianjin port and the waters in the central gulf near the offshore oil drilling platform. In 2003, eutrophied areas of Bohai Bay accounted for 45% of all monitored areas. This is mainly attributed to sewage from land-based sources. In recent years, Bohai Sea has seen more red tides every year, with ever-increasing damage. Additionally, oil spills are frequently. It is estimated that up to 0.2% of the total oil delivery is spilled.

Figure 1.2 Coastal algal blooms in Bohai Sea

From the early 1990s to the early 21st century, the areas of concentrations of DIN and PO_4-P in the surface seawater have expanded from Laizhou Bay, BohaiBay, Liaodong Bay and other coastal waters, into the central parts of the basin. The coastal waters of Bohai Bay are heavily polluted from land-based sewage emission and mariculture development, and which further negatively impact fish and shrimp catches whose economic value has almost disappeared in the estuary areas. Benthos in Bohai Bay has been significantly reduced; only those plankton populations that can survive heavy pollution remain, and the marin eecological environment had been deteriorating. The levels of inorganic nitrogen, inorganic phosphorus, etc. greatly exceed water quality standards, mainly caused by sewage from the coastal rivers flowing into Bohai Bay. Red tides are now frequent. The water quality deterioration in Bohai Sea also has an adverse impact on tourism. One outcome of the decline in freshwater input into Bohai Sea has been a reduction in river pollution loads into the Bohai Bay with the result that water quality has been improving and appears to have stabilized.

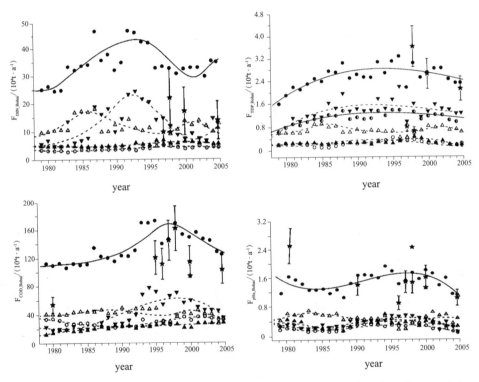

Figure 1.3 Trend of main pollutant flux of flow into from Bohai Bay from the coastal rivers from late 1970s to the early of 21st century. Top left is Dissolved inorganic nitrogen; top right is total dissolved phosphorus; bottom left is COD; bottom right is Petroleum hydrocarbons (o = Hai River; ▲ = Luan River; △ = Liao River; ● = total; ★ = mid-value)

Pollutant flux: Pollutant flux into the Bohai Sea is mainly from land-based emissions through the main rivers around Bohai Sea (Figure 1.3). The proportion of total load coming from the Yellow River and the Liao River is the largest, accounting for about 72%. The proportion total dissolved phosphorus from the Yellow River and the Liao River accounts for about 78% of the total. The proportion of COD from the Yellow River and the Luan River accounts for about 61%. The trend of petroleum hydrocarbons flowing into Bohai Sea is shape of converse "N". into the Bohai Sea have seen a rising, then more recently a falling trend.

Pollution control: Since the 21st century, with the implementation of pollution control plans in Hai River, Liao River Basins, etc., and from the Clean Bohai Sea Action Plan, the environmental pollution of Bohai Sea is improving. Despite large area of polluted waters, at least these are no longer expanding. In summary, the main pollutants that affect larg areas are inorganic nitrogen; active phosphate whereas pollutants such as COD, oil, etc., are found in much smaller areas. Heavy metal pollution is limited to small areas.

1.1.3.4 Ecological Condition in Bohai Sea

Chlorophyll-α concentration in the waters have increased and diversity of phytoplankton has decreased; pollution-tolerance species have become dominant accompanied by a change in the structure of phytoplankton community including dominance of some exotic species. Factors include: rise of sea water temperature, reduced SiO_3-Si/DIN ratio, the gradual rise in salinity and high concentrations of DIN.

Pelagic animal species have increased, accompanied by changesin dominant species that may be related to the shift in their diet of a changing composition of phytoplankton. There was no significant change of the species betweenthe 1950s and the 1980s, while there was a minor increase of species in 2003 compared to the 1980s.

Polychaetes are dominant amongst benthic organisms both in terms of species' composition and abundance; mollusks dominate in terms of biomass and production. There have been significant changes in species' composition with more polychaetes and fewer mollusks and crustaceans. Overall, the biomass of marine benthic fauna has decreased.

1.2 HAI RIVER BASIN

The Hai River Basin (Figure 2) is located between east longitudes from 112 ° to 120 °and north latitude from 35 ° to 43 °; east of Bohai Sea , south to the Yellow River, west to Yunzhong Mountain and Taiyues Mountain and north to the Mongolian Plateau. With a total area of 318,000 km^2 it crosses eight provinces including most of Hebei, the eastern part of Shanxi, northern Shandong and Henan, and small parts of the Inner Mongolia Autonomous Region and Liaoning Province. Autonomous regions and municipalities within the Hai River Basin include Beijing and Tianjin municipalities. There are mainly mountains and plateau in the northern and western parts of the Hai River Basin, and large areas of the North China Plain in the east and south-east. There are 189,400 km^2 of mountains and plateaus, accounting for 60% of the total area, and 128,400 km^2 of plains area, accounting for 40%.

The Hai River Basin contains a large number of rivers; these can be divided into 3 major water systems, the Hai River, Luan River, and Tuhaimajia River. The Hai River can be further divided into the north part (including Jiyun, Chaobai, North Canal, the Yongding River),and the south part (including the

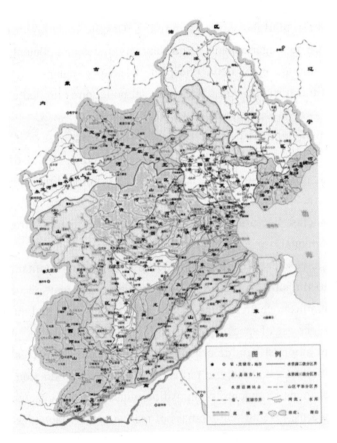

Figure 1.4 Relationship between Hai River Basin, Bohai Sea (A) and Bohai Bay (B) (Source : Water Resource conservation Bureau of Hai River Basin, 2004)

Great River, Ziya River, Zhangweinan River (Canal), the Heilonggang River and Hai River). Most rivers, where they flow into the Bohai Sea are seasonal rivers, with occasional floods during the summer rain season, and very low to nil flowsin winter and spring. More than 70% of the inflow into the Bohai Sea occurs in summer (between July and September). The hydrology of the Hai River Basin is largely artificial with thousands of gates, dams, sluices and other control structures. Rivers are mainly canalized for flood control and for routing water within the basin according to need. There is an average annual runoff of 72 billion cubic meters, with 1.6 billion tons of sediment transported into the sea. There has been a significant decrease in the amount of both flow and sediment into the sea due to the impact of human activities in recent years, which leads to serious degradation in the marine ecosystem and decline in its natural biological service functions.

The coastline of the Hai River Basin accounts for about 40% of the entire Bohai Sea's coastline. The most important regions affected by the influx of pollutants from Hai River Basin into the sea include the Bohai Bay and a small western part of the Liaodong Bay. The relationship between Hai River Basin,the Bohai Sea and Bohai Bay is shown in Figure 1.4.

1.2.1 Socio-economic Profile

Population: There are 26 large and medium-sized municipalities and cities in Hai River Basin, including the 2 municipalities of Beijing and Tianjin. This is China's political, economic and cultural heartland. There are in total 132.58 million people in Hai River Basin in 2004, of which the urban population is 48.1 million, with an urbanization rate of 36%, and population density of 414 persons per square kilometer. There are five cities in the Hai River Basin area that surrounds the Bohai Sea, with a total jurisdiction over 125 counties (districts, cities). These have total population of 29.52million in 2003, accounting about 2.3% of the country's total population. Average population density is 535 persons per square kilometer, which is three times more than the national average population density. The population density of the port city of Tianjin is the largest of all .

Economy: The Hai River Basin is an economically developed area in China and an important primary, manufacturing and high-tech industrial base. It has an important strategic position in the country's economic development. The Bohai Sea Rim area is defined as a broad economic region around the entire Bohai Sea and parts of the coastal areas in the Yellow Sea. The State Council has recently proposed to accelerate the development and opening up of this area, and to bring the development of Tianjin New Coastal Area into the overall national development strategy. The regions surrounding Bohai Sea have become the "engine" for north China's economic development, which was named by economists as China's 3rd economic "growth pole" after the Pearl River Delta and Yangtze River Delta. The Hai River Basin is one of China's agricultural bases and is central to the agricultural economy and food security. Details are provided in Section 1.2.4 of this report.

1.2.2 Environment

1.2.2.1 Water Quantity

The Hai River Basin is located in the temperate sub-humid and semi-arid continental monsoon climate zone. It is cold in winter, with little snow but much wind from the north and northwest.The temperature rises quickly in the spring, with heavy wind, little rain and sometimes sandstorms. It is humid in summer, with much rainfall and wind from the southeast. In autumn the days are crisp with little rainfall.

Since the 1950s , the overall precipitation in Hai River Basin and its total water resources have been decreasing. The basin is affected by climate change, with the average annual precipitation in the basin falling from 560 mm during the period 1956 - 1979 down to 501mm in the period 1980 - 2000 — decrease of around 10.5 %. Average annual precipitation during the period of 2001 - 2007 was only 478 mm, a decrease of 14.6 % compared with the 1956-1979 period.Inter-annual fluctuations within these ranges can be substantial. In 2004 the average annual precipitation was 538 mm which was 0.6 % more than that of the average precipitation between 1956 and 2000.

The consequence of decreased rainfall is that the average surface water quantity in the basin has been reduced from 28.8 billion cubic meters in the 1956-1979 period, to an average of 10.6 billion cubic meters between 2001 and 2007.The average total water quantity (including groundwater)has been reduced from 42.1 billion m^3 to 24.5 billion m^3.These reductions in surface runoff arise both from changes in precipitation and changes in land use. Overdraft of groundwater has resulted in a significant decline in ground water levels and deepening of the unsaturated zone in the soil. Precipitation mainly serves to replenish soil water with a corresponding reduction in surface water production. Measures taken in mountain areas to conserve oil and water, including terrace construction and increasing vegetation cover, has increased water storage capacity. Together with decreased precipitation, these conservation measures have reduced the amount of precipitation that is converted to runoff. Additionally, since the 1980s, there have been more dry years. For 17 of the 21 years between 1980 and 2000, the average precipitation in the basin was less than the long-term average. For the 7 years between 2001 and 2007, there was an average of only 10.6 billion cubic meters, which was only 49% of the average surface water quantity during 1956 - 2000. In comparison with the period 1956 to 2000, these seven years were all dry years.

Decreasing runoff in the Hai River Basin has had significant impacts on the Bohai Sea. Not only is there large fluctuation in inflow into the sea, but there is an overall trend of decreasing inflow. In the wet 1950s, there were an average of 24.1 billion cubic meters inflow, but only 2.22 billion cubic meters in the dry 1980s. The average inflow into the sea was only 1.676 billion cubic meters between 2001 and 2007 whereas in 2004 it was 3.37 billion cubic meters.

There was an average of 23.5 billion cubic meters of shallow groundwater in the Hai River Basin between 1980 and 2000, of which 10.8 billion cubic meters of groundwater was in the mountain areas and 16 billion cubic meters in the plains and intermountain basins, with an additional estimated 3.35 billion m^3 of groundwater between these 2 types of areas (of which 14.1 billion cubic meters in the plain area and 1.95 billion cubic meters in the mountain basins).The average shallow groundwater quantity in Hai River Basin between 1980 and 2000 was reduced by 12.6% compared to that between 1956 and 1979.

In summary, precipitation, surface water quantity, groundwater quantity, and total water quantity in Hai River Basin have all shown an overall downward trend. According to the National Assessment Report on Climate Change jointly issued by the Ministry of Science and Technology, China Meteorological Administration and Chinese Academy of Sciences in 2003, climate warming will be further intensified in the future, and it is predicted that change in the average annual precipitation in North China before 2040 will be from 1% ~ 3%.There will be minor variations in precipitation during this planning period, but it is more probable that there will be a shift to more extreme weather events such as the major storm that inundated the Beijing area in July, 2012. The decrease in water resources caused by climate and land use changes further exacerbate the contradiction between water supply and water demand.

After some 60 years of development, a water supply engineering system has been built in the Hai River Basin which combine the local surface water, groundwater, water from Yellow River, and unconventional water resources. The current total water supply capacity is up to 49.2 billion cubic meters, which provides effective support for the sustainable economic and social development of the basin.

The local surface water engineering system is supported by 36 large reservoirs, 18 large-scale water diversion projects, and supplemented by small and medium-sized reservoirs and diversion works.The water distribution networks cover rural and urban water supply for large and medium-sized cities such as Beijing and Tianjin,Shijiazhuang, Taihang Mountain area, and the grain-producing areas near Yan Mountain,etc.. Annual water supply capacitycan reach 13.9 billioncubic meters.

The groundwater system is mainly composed of deep and shallow wells for groundwater exploitation which includes all exploitable areas in the Hai River Basin. There are currently 1.36 million wells of different types of which 1.22 million are shallow wells with depths less that 120 meters, and 140,000 deep wells with depths >120 meters. The annual groundwater supply capacity is up to 28.5 billion cubic meters. Groundwater is the main source of water supply in most large and medium-sized cities including Beijing. It is also an important source for auxiliary and emergency water use. Groundwater is the main water source for rural living and irrigation in plains areas. In the eastern plain area where shallow groundwater is saline, livelihood and basic irrigation mainly rely on the deep artesian water.

The Yellow River Diversion System ismainly comprised of 27 large and medium-sized sub-projects, such as Panzhuang Trunk Canal, Weishan Trunk Canal , Shengli Channel, etc..Water diverted from Yellow River serves as the main source for urban life and irrigation in northern Shandong, northern Henan, Cangzhou and Hengshui in Hebei province. The annual water supply to the Hai River Basin can reach 5.8 billion m^3. Water diverted from the Yellow River also supplies Tianjin city and some wetlands like Baiyangdian for emergency water use. Water diverted from Yellow River is also source of water supply in cities such as Dezhou, Liaocheng, Binzhou, Changzhou, etc., and is a supplementary water source for the cities of Xinxiang, Hengshui, etc..

In summary, the water supply system that has been built in Hai River Basin is a combination of surface water, groundwater, water diverted from the Yellow River, and unconventional water resources. The total water supply in the basin was up to 36.801 billion m^3 in 2004, of which 11.823 billion m^3 was from surface water, 24.697 billion m^3 from groundwater,4.231 billion m^3 diverted from the Yellow River, and 282 million m^3 from unconventional water sources. Continuing water scarcity has caused the government to build the South to North Water Transfer Scheme in which water is diverted from the Yangtze River northward to the Hai basin. The East Line (following existing canal systems) and the Middle Line of the South-North Water Transfer Scheme, plus the works of Wanjiazhai North Main Line of Yellow River Diversion project, are currently under construction. The average annual water supply from the various engineering works in the basin was about 40 billion cubic meters between 1980 and 2007, which provided effective support for the basin 's economic and social development in terms of water supply.

Evaporation and transpiration (ET) cause water loss from natural processes and human activities. This includes vegetative evapo-transpiration, soil evaporation and surface evaporation. ET, which reflects the absolute loss of water into the atmosphere, is an important indicator for the evaluation and monitoring of water balance. ET is affected by weather, rainfall, land use and agronomic practices, water management and other factors, and is used in this project as an indicator of regional water use and water saving. Water balance calculations involves rainfall, surface water (runoff), groundwater and ET.

From the perspective of the water cycle, water consumption isdivided into two parts. One part is water that is consumed and is not recovered or transferred back into the water balance; the main component of this loss is ET. The second component of the water balance is water that is returned to the hydrological system and includes irrigation return flows, infiltration losses from canals, recycled industrial and municipal wastewater,etc.. From the perspective of water balance in hydrology,with the premise that the ecological water demand such as inflow into the sea is guaranteed,only by reducing ET consumption can the water balance in the basin be realized. The refore, thewater resource management with ET consumption as the core concept starts from the efficiency of water consumption, and focuses on the amount of water consumption in every aspect of the water cycle so that non-beneficial ET (that part that is not used in vegetative growth) is minimized.

1.2.2.2 Water Quality- Ambient Characteristics

China has a five class system of characterizing water quality. Class I is the best quality (drinking water source) and Class V is the worst in which most beneficial uses cannot be supported. In 2007, of 819.3 kilometers evaluated in the river system, 597.4 km (72.7%) was worse than Class III. Between 2006 and 2008, the proportion of >Class V function zones improved (from 33.8% to 25.3% of the total number of function zones), yet the compliance rate of the function zones to required standards remained below 50%. In addition to the conventional pollutants, persistent organic pollutants such as hexachlorobenzene and DDT began to appear. While water pollution was marginally improved the overall situation is not promising. As the Hai River Basin crosses eight provinces pollution control for trans-boundary sections is critical for pollution control in the whole Hai River Basin. Between 2002 and 2007, only 30% of the trans-boundary sections attained the required water quality standard.

1.2.2.3 Water Quality - Pollution Sources

Water pollution in the Hai River Basin is mainly caused by the emissions of industrial and mining wastewater and from urban domestic sewage. A total of 4.75 billion tons (Bt) of wastewater are discharged annually in the basin; 2.63 Bt(55.4%) is industrial wastewater discharge and discharge from the construction industry. Urban wastewater plus the tertiary industry sector account for 28.2% and 16.4% respectively. The main pollutants in the Hai River Basin include Ammonia nitrogen (NH_3-N), BOD_5, oil, potassium permanganate index and volatile phenols. In 2005 these five categories of pollutants accounted for more than 90% of the pollution.

According to environmental statistics, of the 7,741 main polluting enterprises in 2007, there were 2.437 billion tons of industrial wastewater, 641,100 tons of COD, and 51,200 tons of NH_3-N. This was a slight decrease compared to 2005. Pollution is mainly industry-specific. The paper industry and chemical industry are the two largest polluting industries in Hai River Basin and make the largest contributions to total amounts of COD and NH_3-N. In 2007, the total output value of paper and chemical industries accounted for 16.3% of total wastewater, but their contribution to COD and NH_3-N loads was 56.8% and 61.1% respectively.

Chapter 1 OVERVIEW OF HAI RIVER BASIN AND BOHAI SEA

There is also a significant difference in the spatial distribution of industrial pollutant loads. The areas with heavy COD emissions include the provinces of Hebei, Henan and Shandong. Areas with heavy NH_3-N emissions are mainly Hebei and Henan Province. Among the 31cities, Cangzhou, Baoding,Tianjin, Dezhou and Xinxiang contributed the largest amounts to pollutant discharge. By analyzing the industrial pollution indexes such as the wastewater production coefficient , water cycle utilization rate, the pollutant coefficient per unit of output, etc., it was found that the average industrial emission of COD per unit of GDP in Hai River Basin was below that of the national level, and the average industrial emission of NH_3-Nper unit of GDP was the same as the national level. The pollutant load per unit of GDP in Henan and Shandong Province were higher than the national level,and are areas where greater pollution control efforts are needed.

1.2.2.4 Potential for Pollutant Reduction

The sewage treatment rate and water reuse rate in Hai River Basin had been improving but the operation of the sewage treatment plants still needs improvement. In 2005, there was a total of 92 sewage treatment plants with a treatment capacity of 1.74 billion tons, a treatment rate of 52.1%,and a reuse rate of renewable water of 5.1%. In 2007, there were 172 sewage treatment plants with a treatment capacity of 2.63 billion tons, treatment rate of 69.94%, and water reuse rate of 11.06%. About 44% of the construction of urban sewage treatment facilities in Hai River Basin had been completed by 2007. Sixty-nine percent of urban sewage was treated in 2007, with only 8.1% of the sewage treatment plants realizing the design treatment capacity. The COD concentration of treated effluent from 31% of the sewage treatment plants failed to meet the Class I-B Standard, and there was a failure in 49% of the urban sewage treatment plants to meet the Class I-A standard in terms of COD concentration of the treated effluent. The wastewater treatment capacity varies substantially amongst the different regions.

The"11th Five-Year Plan for Water Pollution Control in Hai River Basin", 234 urban sewage treatment projects are to be developed with added treatment capacity of 784.9 tons / day. When these projects are completed and put into operation, and with an operating load rate of more than 75%, COD reduction capacity would be increased by approximately 400,000 tons/year, and ammonia reduction capacity increased by about 30,000 tons/year. It is expected that urban domestic sewage production will reach 5 billion tons/ year by 2020. If the new sewage treatment facilities in the "11th Five-Year

Plan" are completed and put into full operation all the urban wastewater will be treated in Hai River Basin.

As is indicated in the *"National 11th Five-Year Plan for Urban Sewage Treatment and Recycling Facilities"* reused/reclaimed water should be more than 20% of treated wastewater in northern cities where water resources are scarce. If 20% to 40% of the treated wastewater could be used in Hai River Basin, there would be 0.687 to 1.274 billion tons of renewable water reused.

1.2.3 Evapotranspiration (ET)

Evapotranspiration is a process by which water is lost into the atmosphere through evaporation from land and water surfaces, and by transpiration from plants and animals. As noted above, ET is the single largest loss of water and is especially important in this dry basin. ET has a central role in the water cycle, and is the link between vegetation growth and other ecological processes, and hydrological processes. Therefore ET management is a central strategy in conserving available water. It becomes especially important in agriculture which still uses the majority of water in the Hai River Basin and where saving through ET control can be very large.ET's role in the water cycle and energy cycle is illustrated in Figure1.5. The technical description of this strategy is contained in Chapter 3.

ET management is a form of water demand management that focuses on controlling the loss of non-beneficial ET (that part of ET that is not used by plants for growth). ET control is a new concept in Chinese water resource management and is a key issue in this water-scarce basin. For this GEF Hai River Basin IWEM Project, ET management has become a core concept.ET management was originally conceived of within the concept of "Real Water Savings"that was initially introduced by the World Bank in the "Water Saving and IrrigationProject"(WCP) in the period 2001 - 2005.Because the Hai River Basin suffers from water shortage the national and local governments have invested heavily in the development of water-saving irrigation in agriculture. Nevertheless, despite savings produced by more efficient irrigation (engineered water savings), water demand meets supply only be continuing drawdown of groundwater. As this is not sustainable, and to balance water supply and water demand, it requires a fresh look at water savings. Water balance calculations show clearly that the only place where there

Chapter 1 OVERVIEW OF HAI RIVER BASIN AND BOHAI SEA

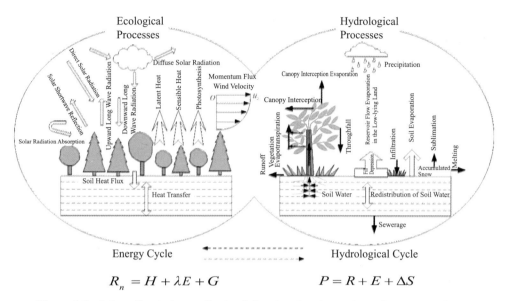

Figure 1.5 Interaction between the basin/ regional water cycle and energy cycle
Notes: In the symbols in the energy cycle : Rn, H , λE and G represent the regional/ basin average net radiation, sensible heat flux, latent heat flux and soil heat flux respectively; in the hydrological cycle, P , E, R , and ΔS represent the regional / basin average precipitation, evapotranspiration, average runoff and river basin water accumulator variable respectively.

remains potential for large savings is in the area of ET control. "Real Water Savings" departs from conventional thinking about engineered water use efficiency. The latter regards, for example, canal losses to groundwater, as losses from the water system. From a water balance perspective, however, this 'lost' water is only transferred from one part of the hydrological cycle to another and, therefore, is not lost. Both perspectives have validity, however from basin-wide perspective, 'real water savings' offers more comprehensive approach to water management in which water in different water cycle compartments (surface water, groundwater, ET, etc.) are managed within holistic framework.

ET management is not a totally new concept for water resource management in China where irrigation water requirement is based on ET of crops. The challenge in this project has been to measure ET with the objective of assigning ET quotas based on transportation requirements of crops and minimization of non-beneficial ET through a variety of agronomic interventions. Under the project, the objective has been to have full

ET management in the basin. As there have been no fully successful examples at home or abroad the Project focused on resolving key technical problems and demonstrating the practical viability of the ET management system.

The key issues developed in the Project for implementing ET management at the basin level are mainly as follows:
- the accuracy of ET data monitored by remote sensing technology;
- validation and coupling of monitored ET data at different scales;
- the distinction between controllable and uncontrollable ET;
- the relationship between ET and manageable water indicators;
- Setting ET targets for different regions.

1.2.4 Social and Economic Situation in Hai River Basin

There are 8 provinces (municipalities/autonomous regions) in Hai River Basin as shown in Table 1.1. Currently there are 57 municipalities (cities) in the basin, including 26 large and medium-sized cities and 31 county-level cities. There are more than 200 counties. The total population in Hai River Basin in 2005 was 134 million, an increase of 38% compared with that in 1980. The total population in the basin accounted for10.2% of the country's total population, of which there were 50.23 million urban population (37.4%), and 83.96 million in the rural areas. The average population density in the basin was (2005) 419 persons/km^2; the population density in the plain area was 747 persons/km^2; and 183persons/km^2 in hilly and mountainous areas(including the inter-basin plain areas). The demographic data for the provinces(municipalities in 2005 are shown in Table1.1).

The Hai River Basin is an important base for industries that have strategic significance for the country's economic development. Major sectors include metallurgy, electric power, chemicals, machinery, electronics, coal, etc.. There has been a rapid development of high-tech industries, such as information technology, biotechnology, green energy, new materials, etc.. The gross domestic product (GDP) in Hai River Basin in 2005 was 2,575 billion Yuan, with a 15-fold increase over 1980 values, with an average annual growth rate of 11.8%. The GDP in Hai River Basin accounted for 14.1% of the national GDP. While GDP per capita was 19,200 Yuan—1.38 times the national average, there is a large disparities in the distribution of regional economic development.

Table 1.2 Demography of Hai River Basin in 2005

Administrative Areas	Total population (in 10,000)	Urban population (in 10,000)	Rural population (in 10,000)	Urbanization ratio/%	Population density/ (per km²)
Beijing	1,538	1,238	300	80.5	915
Tianjin	1,044	667	377	63.9	876
Hebei	6,775	1,803	4,972	26.6	395
Shanxi	1,172	489	683	41.7	198
Henan	1,240	442	798	35.6	809
Shandong	1,523	359	1,164	23.6	492
Inner Mongolia	104	22	82	21.2	83
Liaoning	23	3	20	13.0	135
Basin Total	13,419	5,023	8,396	37.4	419

Table 1.3 Economic development indicators of the Hai River Basin, 2005

Administrative Area	GDP/ ($\times 10^{11}$Yuan)	Industrial Value/ ($\times 10^{11}$Yuan)	Crop acreage/ ($\times 10^4$ mu)	Crop production/ ($\times 10^4$ tons)	GDP per capita/ Yuan	Crop production per capita/ kg
Beijing	6,814	1,782	351	94	44,304	61
Tianjin	3,665	1,885	622	137	35,105	131
Hebei	9,943	4,463	9,078	2,599	14,676	384
Shanxi	1,500	572	2,071	392	12,799	334
Henan	1,627	881	1,071	485	13,121	391
Shandong	2,076	899	2,317	1,000	13,631	657
Inner Mongolia	113	33	445	47	10,865	452
Liaoning	12	2	26	8	5,217	348
Basin Total	25,750	10,517	15,981	4,762	19,189	355

The Hai River Basin is one of the major grain production bases in China. In 2005, there were 159.81 million mu of farmland in the basin, accounting for 33% of the total area in the basin. The effective irrigation areas were 113.14 million mu, and the actually irrigated area was 95.43 million mu, with an irrigation rate of 60%. The main crops include wheat, barley, corn, sorghum, rice, beans, etc.. Cash crops are mainly cotton, oil, hemp, tobacco, etc.. There was a total grain output of 47.62 million tons, accounting for 11.6% of the national output. Average yield was 350 kg per mu. The Taihang Piedmont and Tuhai Majia Plain are the main agricultural areas where food production accounted for 70% of that in the basin. Coastal areas have favorable conditions for the development of the fishery and mudflat aquaculture. Recent years have seen large increase in animal husbandry, fisheries, oil, fruit, aquatic products, meat, eggs, milk, etc..The economic development in Hai River Basin in 2005 is shown in Table 1.2.

1.3 INSTITUTIONAL SITUATION AND LEGAL FRAMEWORK

1.3.1 The Basin Management System

1.3.1.1 Current Policies

Currently the policies that are followed for IWEM in Hai River Basin are as follows :

The macro-management and configuration of water resources: The water planning system takes into account the balance between water supply and water demand, the overall water balance, amounts required for ecological conservation, and practices of water saving. The water plan involving water development and utilization is based on coordination amongst these components and allocations are assigned. The water withdrawal permitting system is implemented and charges are levied for water use.

In the future the water permitting system will be strengthened with the expansion of water metering. Water resources assessment will be strengthened, and more specific requirements will be made for water withdrawal for construction projects, improved rationalization of water withdrawal and water return(discharge), the impact on the water environment and on other water users. Water allocation plans and water regulation plans in cases of emergency situations (drought) will be improved as well as the annual water allocation scheme and scheduling.

Water resource conservation and management: The planning system is made for water resource conservation, to strengthen the management of water function zoning, and to clarify water environmental protection objectives and the pollutant assimilative capacity. With a system with both control of total water use and quota management, an annual water use plan is developed, and the control of total quantity is implemented. Quota management is implemented for water users and extra charges are made for water use that exceed their quota. Drinking water source conservation areas are delineated and protected. The corresponding requirements for construction projects are made clear, and appropriate compensation mechanisms for ecological protection are developed. Management information systems will be improved for water function areas and the quality of water in these function zones will be regularly put in the public domain so that public's right to be informed is ensured. Systems are to be improved of setting outfall limits and for the control of total emission of pollutants (total load control), with strict procedures for application and review of setting pollutant objectives with the objective of effective control of pollutant emissions.

Water pollution prevention and control, and water environmental management: Responsibilities are set for environmental protection objectives and an appraisal system identifies responsible entities and their respective environmental protection obligations. Water pollution control planning system is implemented as the fundamental basis for water pollution control; a supervision and inspection system ensures appropriate implementation. Sewage discharge permitting and registration system regulates sewage dischargers, and to ensure rigorous pollution discharge declarations, approved sewage discharge volumes,licensing, and administrative procedures to deal with excessive discharges. A charging system is in place for pollutant discharge; charges are made according to the emission types and quantities of dischargers having direct discharges into water bodies. There is a water environmental quality monitoring and water pollutant emission monitoring and unified information dissemination. Pollution control measures are required to be implemented before deadlines, and explicit information is circulated about penalties for excessive emissions or emissions with pollutant concentrations that exceeding the standards. Lists are made for production processes and equipment that are substandard and are to be eliminated within specific deadline; supervision of producers, sellers,importers and users is enforced in order to stop the production, sales,import and use of devices and processes that are to be eliminated. Control measures are taken for the total load control of key water pollutants, with indicators set for the reduction of key water pollutants at the provincial and lower levels. A response system is established for emergency pollution incidents, with clear implementation procedures,quick response mechanisms and the necessary material reserves in place for emergencies. There is a disclosure system for environmental information, with public notification of environmental protection policies and regulations, project approval process and results, etc.. The business section's environmental information is also required to be open to the public.

1.3.1.2 The Existing Regulatory Framework

National level: There are 8 national laws including the Water Law, and the Environmental Protection Law; 15 administrative regulations including the Hydrology Regulations; 6 normative documents issued by the State Council, including " The State Council's decision in forwarding the notification by National Environmental Protection Administration and Other Departments to Strengthen Water Environmental Protection in Key Lakes"; 28 rules and regulations by the State Council such as the Interim Measures of water allocation, and 38 departmental normative documents such as Regulations On Water Functional Zone Management.

At the basin level: There are 16 normative documents, such as "Clarification of responsibilities and Administrative jurisdictions of HWCC" and "Notification on the Implementation of Responsibility-specific System for Water Engineering Projects".

At Local Level: There are more than 50 local administrative regulations and rules concerning the water resource and water environmental management in Hai River Basin, as well as 58 local government departmental rules and 84 pieces of local normative documents.

1.3.1.3 Institutional Setting

National Level: The institutions involved in this project, and which have the primary responsibility for water and environment are the Ministry of Water Resources (MWR) and the Ministry of Environmental Protection (MEP). MWR is responsible for water resource development, utilization, conservation and management, with the combination of basin management and administrative area management. MEP is responsible for the water environment quality management and water pollution control through its network of subordinate Environmental Protection Departments (provincial) and Environmental Protection Bureaus (EPBs) at lower levels, and with the cooperation of other relevant agencies horizontally and vertically. The management system is to ensure that governments at all levels are responsible for the environmental quality. In addition, the State Oceanic Administration (SOA), the National Maritime Authority and the National Bureau of Fisheries are also involved, especially in work involving the Bohai Sea.

Basin level: This mainly involves the Haihe (Hai River) Water Conservancy Commission (HWCC), North China Environmental Protection Inspection Center, and local governments and provincial and lower levels. The HWCC contains five departments: Planning and Programming Department, Department of Water Administration and Water Resources,Construction and Management Office, Soil and Water Conservation Department, and Flood Control and Drought Relief Office. There are 5 subordinate institutions under the charge of HWCC: the Zhangweinan Canal Administration Bureau, the Downstream Administration Bureau, the Upstream Administration Bureau,the Hydrology Bureau ,and the Soil and Water Conservation Monitoring Centre. Located within the HWCC there is the independent Water Resource Protection Bureau (WRPB) of Hai River Basin with particular responsibilities for transboundary monitoring,and the subsidiary of Luan River Diversion Project Authority. Together, these subsidiary units are responsible for the legislation and its enforcement in the basin,for basin planning and implementation, and for integrated water resources management in the basin.

The North China Environmental Protection Supervision Center is the supervising agency for law enforcement for the Ministry of Environmental Protection. On behalf of the Ministry of Environmental Protection, it is responsible for supervising the enforcement of local and national environmental policies, plans, regulations, and standards in the provinces under its jurisdiction. It is also responsible for the coordination of inter-provincial (regional/basin) environmental disputes. Local legal departments are responsible for the drafting of local laws and regulations on water management,and local governments are responsible for mid and long-term development planning and annual plan for water resource development, as well as coordination and management for the regional water conservation management.

Local level: This mainly involves agencies of local governments that are responsible for water conservancy,environmental protection,land,construction, transportation, agriculture, etc.. Within this project local governments in the 16 pilot areas of the basin made significant achievements in water resource management by promoting integrated water resource management.

1.3.2 Analysis of the Problems of the Present Management System

The present management system has evolved over many years with water quantity under the management of MWR (this institution predates modern China), and environment (including water quality) under the jurisdiction of the much younger and smaller MEP. Because the Chinese institutional system is a "command and control" system and is vertically structured, the interaction between MWR and MEP has been problematic for many years and has been frequently characterized by disputes over mandate, especially in the area of water quality and water quality management where both institutions play a role. This lack of effective horizontal coordination was a major factor in the rationale for this project which had, as its principal institutional objective, an effective integrated management system for water and environment at all levels in the basin. The institutional objective was to develop the policies, coordination mechanisms and joint technical interventions that would demonstrate to the rest of China how IWEM can be implemented. This institutional objective was implemented at national (MWR, MEP), basin (HWCC) and local levels in four provinces and in 16 pilot counties.

1.3.2.1 Policy Issues

- There is no effective combination of water withdrawal, water consumption and water drainage management. It is hard to guarantee the quality and quantity of the water retreat to the downstream, and there is a lack of effective convergence between pollution source management and water protection.
- There are deficiencies in the overall planning for the development of water resource and water environment monitoring network; the site layout is irrational; the monitoring methods and application standards are inconsistent.
- The basin management system results in overlaps, contradictions and gaps between basin management and administrative regional management, with a lack of effective implementation and management of cross-boundary watershed management.
- There are insufficient "workable" regulatory policies, weak accountability policies and administrative regulatory mechanisms, and inadequate mechanisms for public involvement.

1.3.2.2 Legal Issues

- There are no specific provisions in Chinese water or environmental legislation for integrated water resource and water environment management (IWEMP). Because the Water Law and Water Pollution Control Law do not contain provisions for Integrated Water and Environment Plans (IWEMPs) the legal status of IWEMP remains unclear.

- There are no provisions for water allocation control based on ET. In the existing laws and regulations, water allocation is only made among administrative regions at different levels based on the total usable water resources or total exploitable groundwater, with no integration of ET concept into water allocation in the basin. Again, the legal basis for introducing ET as a control methodology is unclear.
- There are no specifications for a unified regulatory body for water conservation and water pollution control in the basin. The current Water Law is mainly focused on water quantity management in the basin and the HWCC has no jurisdiction over pollution control. Therefore, the supervision and management of water resource protection is fragmented.
- The integration of water conservation and water pollution control should be enhanced. The current laws do not clarify the relationship between water resource protection planning and water pollution prevention planning, especially when it comes to the control of total discharge into the river, the assessment of environmental impact, and the systems for land-based pollution control.
- Water withdrawal permitting and discharge permitting are not integrated. Water intake permitting only takes into account the location and quantity of water withdrawal while ignoring water withdrawal quality; pollutant emission permits are issued separately after reviewing the major pollutant emissions into the water bodies in administrative areas.
- The provisions for groundwater conservation are weak. Basin management authorities are unable to exercise complete control on the groundwater well drilling and groundwater extraction, and are not able to implement a conjunctive approach to the allocation of surface water and groundwater in order to optimize the allocation of water resources in the basin.
- There are overlaps in the relevant provisions of water resource monitoring. The agencies of water conservancy, environmental protection, land, agriculture, etc., carry out monitoring work for their own needs but there is little data sharing and no effective mechanism for the unified planning of the monitoring system. These results in redundant construction of monitoring stations and inconsistent monitoring methods and standards.
- Supervision and management need to be implemented for the cross-boundary water

quality. The Water Pollution Control Act, as amended in 2008, gave the water resource protection agencies in the basin the mandate to supervise and inspect water quality at provincial cross-boundary areas. This supplements the existing Water Law but is not fully implemented.

1.3.2.3 Institutional Issues

Inter-organizational coordination mechanisms are lacking at the national level between water resource and environmental management ministries and their respective departments. Under the Water Law, MWR proposes limits for sewage discharge in order to meet ambient water quality targets in rivers and lakes, however MEP usually ignores these recommendations and establish their own plans for regional water pollution prevention and control, plans for the total pollutant discharge control and the issuance of emission permits. This results in serious gap between the management of water resources and water environmental management.

The institutional framework for IWEM at the basin level is imperfect, mainly in the following aspects:

- The decision-making and coordination of basin management agencies are not sufficiently effective. Firstly, the basin management agencies are under the direction of the water administrative department of the State Council, which can only be responsible for issues decentralized by MWR; secondly, there is no participation in the basin management institutions (HWCC in this case) of representatives and stakeholders from administrative regions in the basin; thirdly, the river basin management institutions are relatively weak in their own management capacity.

- The water environment management in the basin is mainly based on administrative regions. It has no representation in the HWCC, and there is little overall planning, coordination, and regulation at the basin level. The HWCC Water Resources Protection Bureau is only responsible for the monitoring and publication of water quality information and for planning for water resource protection, and has no jurisdiction for law enforcement and supervision of the basin water quality conservation. The WRPB is part of MRW and has little effective linkage with MEP. The North China Environmental Inspection Center is only effective for inspection responsibilities, and weak in decision-making and coordination at the basin level.

- There is no platform at the basin level for the communication and consultation

between water conservancy agencies and environmental protection agencies. The dual leadership of the Ministry of Water Resources and Ministry of Environmental Protection over the Hai River Water Resource Conservation Bureau has been transformed to the unilateral leadership by the Ministry of Water Resources, with communication with the environmental protection agencies missing. Therefore a joint mechanism for water resource protection and water pollution control is lacking.
- Unlike local water bureaus that, while reporting to local government, have strong linkages with the basin water administration, environmental protection bureaus have no basin level support and, while taking direction from MEP, tend to be strongly dependent on local government.Because local government tends to put economic growth first and water protection second, it is often difficult for local EPBs to enforce pollution laws.
- There is no provision for public participation in the decision-making for IWEM in the basin. The combination "top down" and "bottom-up" management system has not been put into practice, and it is difficult for the public to participate in decision-making process for IWEM in the basin.
- The responsibilities and authorities are not clarified for basin management agencies and water administration departments of the local government. There are no clear combinations between the watershed management and administrative area management.
- Integrated water resource management has not yet been realized in parts of the basin areas, which result in the fact that water source areas are not responsible for water supply, water supplying agencies are not responsible for water drainage, and water drainage management agencies are in not in charge of water reuse. These disconnects in management practices make IWEM difficult to implementin the basin.

1.3.2.4 Summary of Key Management Issues that Inhibit IWRM

IWEM is accepted as an effective way of taking into account both water resources and water environment at the basin level, and is the basis for establishing an integrated management mechanism to resolve the conflict between these two areas. It puts an emphasis on the integration of basic information and decision-making processes, and aims to make comprehensive and integrated arrangements for water management, pollution control, and water environment protection in the basin. The principal reasons for the disconnect between water and environment are:

- Separation, legally and institutionally, of water resources and environmental management.
- Lack of effective coordinating mechanisms between these two area at national, basin and local levels.
- Vertical management systems that focus on ministry powers results in each ministry focusing on their own interests.
- Lack of clear policies and regulations about the integration of water quantity and water quality management, pollution source management and protection of water bodies.
- No integration between water resource assessment, water intake license management, and supervision and management of water function zones, and environmental impact assessment and discharge permit management.
- Deficiencies in planning and management systems within ministries. Some key deficiencies include: lack of operational measures such as inter-provincial water allocation plans, upstream and downstream ecological compensation mechanisms; lack of binding requirements on trans-boundary water quantity and quality; inadequate linkages between basin and administrative regional management systems; overlaps and contradictions among the water planning in administrative regions.

1.4 PLANNING SYSTEM IN HAI RIVER BASIN

1.4.1 Plan Types

There are three levels of water-related plans in the Hai River Basin:(i) comprehensive plans, (ii) sectoral plans and (iii) special plans. Comprehensive plans focus on the basin as a whole. These include the Comprehensive Planning in Hai River Basin, the "Five-Year" Plan for Water Resource Development in Hai River Basin, etc.. These contain all aspects of water-related issues such as water use, water conservation, aquatic ecosystem restoration, water and soil conservation, flood control, shoreline use, information management, integrated management, etc..Sectoral plans focus on some specific area and are guided by comprehensive plans. Examples include Water Resources Planning, the Five-Year Plan for Water Pollution Control, the Flood Control Plan, Plan to Restore Ecological Environment, etc.. Special plans are time-limited and are created to solve specific issues or issues within a local area or river. Examples include Opinions on the Recent Flood Control Construction, Plan for the Sustainable Water Use in the Capital at early 21st century(2001-2005), the Master Plan for South-to-North Water Transfer Project, and comprehensive plans for a local rivers or river systems.

1.4.2 Planning Procedures

1.4.2.1 Authorization of Planning

The Department of Water Administration (Environment) of the State Council issues planning tasks to the basin management organizations according to the Planning Document approved by the State Council. The basin management organizations then organize the water administrative departments of the provinces (autonomous regions and municipalities) within the basin to carry out specific planning, and to form and submit the basin planning report. This is approved by the Water Administration Department of the State Council after review.

1.4.2.2 Principles in Planning

The principles behind planning are to build a resource-conserving and environmentally friendly society, sustainable water development, and to promote harmony with nature and maintain river health and sustainable use.Therefore, plans attempt to rationally develop water resources, protection against floods and droughts,integrate the basin management to the extent possible, achieve pollution reduction, control costs, promote river basin development and reform, promote economic development and achieve water use efficiencies, and to provide water and environment services to the public.

Important principles that are not now well represented in planning are:
- Scientific water management with focus on management innovation.
- Relationship between the basin and administrative areas, and the linkage between upstream and downstream, and with both sides along the rivers.
- A coordinated approach to industries in the basin to realize sustainable development in the basin.

- A consultation system for basin management.

1.4.2.3 Specific Planning Objectives

Overall objectives include:

Total water use: The total water use will reach 49.5 billion cubic meters by the year of 2020, and 50.5 billion cubic meters by 2030.

Security of water supply security: By the year of 2020 water supply will be increased by 8.2 billion cubic meters, with water supply in key areas guaranteed. By the year of 2030, water supply capacity will increase by 13.2 billion cubic meters, with urban and rural areas significant drought-proofed. Also, by 2030 urban and rural water supply security system will be implemented.

Water-saving: By the year of 2020, the penetration of domestic water-saving appliances and industrial water reuse rate will reach 95% and 87% respectively; the irrigation water utilization coefficient will be 0.73; the water consumption per 10,000 Yuan of GDP will be reduced to 69 cubic meters; the water consumption for every 10,000 Yuan of industrial added value will be reduced to 29 cubic meters;ET in the entire basin will be around181.9 billion cubic meters.

By the year of 2030, the penetrationof domestic water-saving appliances and industrial water reuse rate will reach 99% and 90% respectively; the irrigation water utilization coefficient will be 0.75; the water consumption per 10,000 Yuan of GDP will be reduced to 38 cubic meters; the water consumption for every 10,000 Yuan of industrial added value will be reduced to 17cubic meters;ET in the entire basin will be around 182.7 billion cubic meters.

Surface water conservation: By the year of 2020,all the water source areas for urban water supply in the basin will be in full compliance with water quality standards; the total discharge of COD into the river will be within530,000 tons, and the total amount of ammonia into the river will be within 50,000 tons; up to 63% of the water functional zones will reach water quality standards; effective systems will be built for water resource protection in the basin. By the year of 2030,the total COD discharged into the river will be controlled at 310,000 tons, and the amount of ammonia into the river will controlled at 15,000 tons; all of the water functional zones will reach water quality standards.

Groundwater resource protection: By the year of 2020, overexploitation of shallow groundwater will be reduced by more than 50%, and a withdrawal-recharge balance will be realized by 2030. Water quality should not be inferior to the quality of current case. The continual and rapid decline of the groundwater table in over-exploited areas will be gradually curbed, and the problems of sea water intrusion, salt water intrusion, ground subsidence, spring flow attenuation, groundwater pollution and other ecological and environmental problems, which were caused by overexploitation of groundwater, will be gradually brought under control.

Aquatic ecosystem protection and restoration: A security system for ecological and environmental water use will be established; the ecological water quantity in the rivers will be no less than one billion cubic meters in key mountain areas, and no less than 2.8 billion cubic meters in the main rivers in plain areas. By the year of 2020, inflow into the sea will be no less than 6 billion cubic meters; the excessive extraction of groundwater will be controlled within 4 billion cubic meters in plain areas. By 2030, inflow into the sea will be no less than 6.5 billion cubic meters, and a balance will be realized between groundwater withdrawal and recharge.

Objectives for soil and water conservation: By the year of 2020, another 60,000 km^2 of areas with soil erosion will be brought under control, with a total of more than 70% of affected areas controlled; a more advanced system will be put into place for the supervision and dynamic monitoring of soil and water conservation. By the year of 2030, another 20,000 km^2 of areas with soil erosion will be brought under control, with a control rate of more than 90%; a comprehensive supervision and monitoring system for soil and water conservation will be implemented.

Flood control: By the year of 2020, all the Grade 1 and Grade 2 dikes in the basin will reach the appropriate standard; the key flood storage and detention basin will be put into safe use; the downstream areas and major cities in the basin will meet the national flood control standards. By 2030, all the areas will meet the flood control standards, and a modern flood control system will be accomplished.

Chapter 2
PROJECT CONTENT AND TECHNOLOGY ROADMAP

2.1 THE GEF HAI RIVER PROJECT

The Hai River Basin Water Resources and Water Environment Management Project is a large scale project under OP 10 of the International Water Portfolio of the GEF. It is jointly implemented by the Ministry of Finance, the Ministry of Water Resources and Ministry of Environmental Protection with the World Bank acting as the executing agency.Following a preparatory phase that began in 2002, the project passed a formal assessment of the World Bank delegation on December 6, 2003. On April 15, 2004, the GEF Council approved the *China Hai River Basin Water Resources and Water Environment Management Project*, which came into effect on September 22, 2004.

The GEF contribution was USD $17 million with domestic co-financing of $16,320,000. This project was led by the Ministry of Finance, and implemented by the Ministry of Water Resources (MWR), the Ministry of Environmental Protection (MEP), with implementing authority delegated downwards to the IOC of Zhangweinan Canal sub-basin, and water conservancy and environmental protection departments in Tianjin, Beijing, Hebei Province and 16 pilot countries. The GEF Hai River project requires the greatest degree of horizontal and vertical integration and cooperation at the core. Horizontal integration includes inter-departmental cooperation such as cooperation between water conservancy and environmental protection departments at all levels, and coordination with the agriculture and construction sectors; vertical integration refers to efficiency management arrangements and coordination within each of the key sectors at the central level and Hai River Basin agencies, Zhangweinan Canal Authority, and key counties (cities, districts)in Tianjin, Beijing and Hebei Province.

2.2 PROJECT OBJECTIVES

2.2.1 Geographical Scope

The scope of the project is the entire Hai River Basin, with longitude 112° ~ 120°, latitude 35° ~ 43° and total area of 318,000 km². It includes Beijing and Tianjin and most part of Hebei Province, the eastern part of Shanxi, Henan, the northern part of Shandong, and a small part of Inner Mongolia Autonomous Region and Liaoning Province. However, the project focused especially on pilot and demonstration work in 16 key counties (districts, cities) in which IWEM planning was carried out (Beijing, Hebei, Tianjin, Shanxi, Henan, Shandong) as well as several more narrowly focused work on pollution control in Dagu Canal and Binhai Coastal Area of Tianjin.

Table 2.1　IWEM Project areas　Yellow highlighted areas are 16 pilot counties and districts

PROVINCE / REGION	COUNTY/DISTRICT	TARGET IWEM ISSUES
Beijing Administrative Region	Tongzhou District	Wastewater Reuse Sustainable groundwater use, water rights, well permitting Water Savings & High Efficiency Water Use
	Daxing District	Sustainable groundwater use, water rights, well permitting Water Savings & High Efficiency Water Use
	Fangshan District	Sustainable groundwater use, water rights, well permitting Water Savings & High Efficiency Water Use
	Miyun County	Sustainable groundwater use, water rights, well permitting Water Savings & High Efficiency Water Use
	Pinggu District	Sustainable groundwater use, water rights, well permitting Water Savings & High Efficiency Water Use
Tianjin Administrative Region		Wastewater Reuse Water Pollution & water supply Groundwater quality & quantity Wastewater management

PROVINCE / REGION	COUNTY/DISTRICT	TARGET IWEM ISSUES
Tianjin Administrative Region	Hangu District	Pollution control
	Ninghe County	Agricultural non-point source pollution
	Baodi County	Pollution control
	Dagu Canal (Tianjin coastal area)	Contaminated sediments Wastewater treatment
	Binhai Coastal Area	Wastewater treatment
Hebei Province	Shijiazhuang City Baoding City Tangshan City	Wastewater Reuse
	Guantao County	Surface and ground water pollution control Sustainable groundwater use, water rights, well permitting Water Savings & High Efficiency Water Use
	Cheng'an County	Sustainable groundwater use, water rights, well permitting Water Savings & High Efficiency Water Use
	Linzhang County	Water scarcity
	Feixiang County	Pollution control
	Shexian County	Sustainable groundwater use Pollution control
Henan Province	Xinxiang City	Wastewater Reuse
	Xinxiang County	Pollution control
Shandong Province	Dezhou City	Wastewater Reuse Aquatic ecological rehabilitation
Shanxi Province	Datong City	Wastewater reuse
	Lucheng County	Pollution control
Zhangweinan Sub-basin	All four provinces	Water pollution control

2.2.2 Project Goals

The objective of Basin Water Resource and Environment Management (IWEM) is to address the gap in IWEM in China by developing, then applying an integrated approach to institutional management of water and environment with the objective of rationally allocating water resources; to promote efficient and effective water utilization; to restore the ecological environment; to develop planning procedures that will have the effect of alleviating water shortages; to reduce land-based pollution to the Bohai Sea; and to significantly improve the water environment of the Hai River Basin and the Bohai Sea.

2.2.3 Specific Objectives

The specific objectives of the GEF Hai River project were:

(i) To promote and establish water resources and environment management systems suitable for the characteristics of the Hai River Basin, including:
- water resources utilization based on the amount of water resources and on water environmental capacity
- water resources management system based on the theory of water rights
- the implementation of water pollution control
- establishment of upstream and downstream compensation policy
- promotion of water pricing policies based on demand management
- implementation of pollution permits
- strengthening groundwater management water conservation
- strengthening wastewater reuse

(ii) To demonstrate the utilization of integrated water resources and water pollution control in key areas, including the encouragement of industrial restructuring measures, demonstration sewage treatment plants of small cities, wastewater treatment and reuse, agricultural restructuring and sustainable use of groundwater management.

(iii) To strengthen knowledge management (KM) and capacity-building of the Hai River Basin, and to construct a remote monitoring center for ET management across the basin.

(iv) To carry out 8 strategic research programs at the national level, the Hai River Basin level and the Beijing city level and to provide measures to strengthen the integrated management of water resources and water environment of the basin.

(v) To launch demonstration pilots of "integrated water and environment management plans" (IWEMPs) in key counties (districts) of Bejing and Hebei, Tianjin and all their subordinate counties (districts).

(vi) To assist in the design, construction and operations of new municipal sewage treatment plants in small towns of Tianjin Binhai New Area that directly discharge into the Bohai Sea.

(vii) To build a knowledge system based on integrating databases, models and decision support systems to assist the IWEM process in all areas of the project.

(viii) Support the formulation of policies, regulations and institutional framework related to water resources and water environment management on the national level, river basin level, sub-basin level and provincial (city) level.

2.3 OVERALL PROJECT TECHNICAL FRAMEWORK

Figure 2.1 describes the technical outputs of the GEF Hai River project and the relationship between each other.

2.4 MANAGEMENT OF THE PROJECT

Integrated management of water resources and water environment is a new concept that is the heart of this project and requires the unified management of water quality and quantity. To this end, the project established new procedures for water resources and water environment management agencies at local, municipal and basin level and provided policy assistance to organizations at all levels.

City and county level: Governments at all levels set up coordinating bodies that brought together the relevant institutions that manage water resources and water environment. This was facilitated by a project office in each pilot area.

Sub-basin level: A project office was established in the Zhangweinan canal sub-basin.

Basin level: A project office was established in the Hai River Water Conservancy Commission.

Central level: A senior-level coordination committee was comprised of a number of departments. The duties of this office were mainly coordination; the main work of project implementation was devolved to the CPMOs (below). Departments that participated in the implementation of the project were: Ministry of Water Resources, Ministry of Environmental Protection, Ministry of Construction, State Oceanic Administration, Ministry of Agriculture, Ministry of Finance, Beijing, Tianjin, Hebei Province, the Hai River Water Conservancy Commission. The executing agency of the project is the Ministry of Water Resources and the Ministry of Environmental Protection at the central level, and province-level institutions Beijing, Tianjin and Hebei.

For practical purposes, the project components were assigned to one or other of

Chapter 2 PROJECT CONTENT AND TECHNOLOGY ROADMAP

MWR and MEP for day-to-day management purposes. Each ministry established a Central Project Management Office (CPMO) for this purpose. The CPMOs were mainly responsible for project implementation, including performance management, contract issuance and supervision, financing and progress reporting, and direct supervision of the work of the project. The two CPMOs worked in a highly interactive manner.

2.5 TECHNICAL QUALITY

Technical quality control was achieved using several levels of inputs:

International Expert Panel (IEP): A number of well-known, senior, international consultants were contracted by the CPMO on the recommendation of the World Bank. The experts represented the major focus areas of the project(water quality, water management, water resources, groundwater, water and environmental planning,knowledge management (KM), and remote sensing-ET management),and for the most part, had extensive prior experience in China. The members of the IAP came to China several times each year to elaborate on technical issues, carry out training, review technical progress, comment on reports and assist the CPMOs with some areas of technical management of the project.

Central Groups of Experts: The CPMOs of MWR and MEP each established their own senior expert group to provide thematic guidance and depth of management experience for those areas for which each office was responsible. From these two groups, MWR and MEP jointly established the Central Joint Group of Experts (CJGE), whose main task was to provide independent advice on the progress and technical content of the project as a whole.

Central Independent Audit Group: To strengthen the inspection and supervision of the GEF Hai River project, and based on legal requirements of the Chinese Government, the Central Project Office established a central group of national independent experts from MWR and MEP whose role was to carry out two independent audits of program performance each year, to prepare an independent technical report based on their inspection and evaluation of the effectiveness of project management, project coordination, and activities. This independent group also inspected the central project coordination group, the GEF Hai River project direction committee of Ministry of Water Resources and Ministry of Environment Protection, and the World Bank

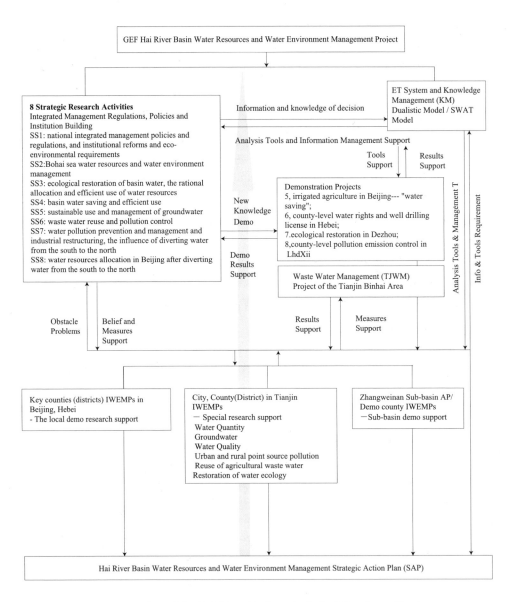

Figure 2.1 The GEF Hai River Project Technical Roadmap Design

Inspection Delegation. Their reports are required by the government as independent verification of the project's progress and performance.

2.6 PEFORMANCE MANAGEMENT

From the beginning of the project a system of Monitoring and Evaluation was established to ensure that the outputs and outcomes identified in the Project Document were met in a timely manner and with appropriate quality. The system included both top-down and horizontally integrated management mechanisms and a bottom-up reporting system. There was a composite technical indicator system to monitor the qualitative and quantitative implementation of the project.

2.6.1 Project Monitoring and Evaluation Management System

M&E is a requirement of GEF projects implemented by the World Bank. A full M&E performance management system was established at all levels, led by the M&E group of the Central Project Office. The composition of the management of the GEF Hai River project monitoring and evaluation is noted in Figure 2.2. Each level was

Figure 2.2   GEF Hai Project River Monitoring and Evaluation Management Organizational Structure

responsible for upwards reporting, and the lower level reports were abstracted to provide M&E evaluation for the entire project.

2.6.2 Indicator System for Project Monitoring and Evaluation

The project monitoring and evaluation indicator system is based on the global "impact/output indicators" identified in the Project Appraisal Document (PAD) at the beginning of the project. These were:

(i) Establishment of a functioning inter-agency committee at the county level, resulting in improved cooperation and integration of Water Resource Management (WRM) and pollution control activities with support from upper levels (prefectures, provinces, Hai Basin Commission (HBC),Zhangweinan, Ministry of Water Resources (MWR), and State Environmental Protection Administration (SEPA).

(ii) Achieve the adoption of improved WRM and pollution control approaches at the county level by institutions implementing Integrated Water and Environment Management (IWEM), including evapotranspiration (ET) management and knowledge management (KM), water rights and well permit administration, and discharge control, with support from upper levels (prefectures, provinces,HBC, Zhangweinan, MWR and SEPA).

(iii) Implementation of improved small city wastewater management approaches in Tianjin coastal counties, including collection, industrial pre-treatment, wastewater treatment, and wastewater reuse.

(iv) Reduce discharge pollution load by 10% in pilot counties and coastal counties.

(v) Reduce groundwater overdraft for irrigation purposed by 10% in pilot counties.

(vi) Reduction of pollution loading to Bohai Sea from at least one Tianjin small city by 10,000 tons of COD and 500 tons of NH_4^+ annually.

(vii) Disposal of 2.2 million cubic meters of contaminated sediment from the Dagu canal in an environmentally safe manner, and achieve a one-time reduction of 10,000 tons of oil, 2,000 tons of zinc, and 5,000 tons of total nitrogen.

The content of these impact/output indicators is highly compressed and required that the project break them into sub-indicators so that each project unit could report its contribution to the larger indicators. Impact indicators and process indicators were used; impact indicators are used to reflect the achievement of the environmental and institutional outcomes expected in the project. Process indicators reflect the implementation targets scheduled throughout the project and are generally shown as a "percentage of completion".The procedure for M&E reporting across the entire Project is shown in Table 2.2.

2.7 CAPACITY BUILDING

Capacity building included domestic and international consulting, technical assistance, operational management assistance, training in wide variety of topics, national and international workshops and conferences, and supervision and inspection (by the World Bank and the International Expert Panel). This was focused both on technical areas and on project management capacity. Capacity development also included enhanced capacity in hardware and software systems as basis to make more informed decisions on water and environmental management, and as basis to share information amongst the various participants in the project. The M&E process was developed around Management Information System (MIS).

2.7.1 Top-Down, Bottom-Up and Horizontal Management Principles

Management of project by two ministries is not usual in China due to the vertical structure of the government system. Therefore, the two ministries jointly formulated and implemented the GEF Hai River Project Cooperation Mechanism. This included horizontal cooperation across ministries and their subordinate agencies at all levels, and vertical integration to ensure that there was systematic management at all levels within each ministry. Additionally, as part of the development of water and environment management plans it was essential to build a "bottom-up" approach in which local stakeholders (citizens, local agencies etc.) had major input into the final content and design of IWEM plans. The project facilitated much bottom-up capacity development so that stakeholders could claim "ownership" of the final product and of the successes achieved in the geographical areas (e.g. counties) under their control.

2.7.2 Training

Project technical training strengthened the abilities of officials and technical staff in the integrated management of water resources and water environment. Study tours broadened the horizons of the project officials and technical staff. Training was focused in the following areas:

- Project Management: finance, accounts, reimbursement payment, bidding, and procurement(all levels within the Project).
- M&E: training of M&E staff at all levels.
- Professional Development: effective presentations, report writing, etc..
- Technical Training: this encompassed all technical areas of the project-knowledge management, ET measurement and planning, IWEM planning processes; groundwater management, pollution control, water rights, market-based approaches to water management, non-point source pollution measurement and management, Community-Driven Development (CDD) and farmer Water User Associations (WUAs), etc..

2.7.3 Baseline Survey

2.7.3.1 Purpose and Scope of the Baseline Survey

The baseline survey was carried in the following areas: the 16 project counties (cities, districts),Tianjin, Tianjin Construction Committee TUDEP, the Hai River Basin, and the Zhangweinan Canal Sub-basin. The survey baseline year was 2004 and included demographic data, pollution loads, water discharge to the Bohai Sea, groundwater withdrawal rates and amounts, ET quantity, the institutional situation, etc.. The purpose of the Baseline Survey was to (a) establish realistic objectives for the project and (b) as a basis to establish the project monitoring and evaluation system that was used to assess the real progress of the project in achieving pollution reduction, real water savings, etc.. Additionally, the baseline survey provided basic data to populate the knowledge platform used in the preparation of IWEM plans (IWEMPs) and Strategic Action Plans (SAPs) for the project counties (cities, districts), in some strategic studies and some related case studies. The basic sources of data of the baseline survey are shown in Table 2.3.

2.7.3.2 Content and Reporting of the Baselines Survey

The baseline survey was conducted according to standard templates and workplans. Similarly the baseline report was according to criteria provided by the World Bank and a workplan developed by the Project. Content of the baseline assessment included the following main areas of interest:

Socio-economic data: Population, Gross Domestic Product (GDP), social / domestic water consumption per capita, irrigation water consumption, irrigation water consumption per mu, industrial water consumption, water consumption for each 10,000 yuan production, irrigation water utilization coefficient, canal water utilization coefficient, industrial water recycling rate, urban pipe water loss rate. Additionally, data such as surface water (rivers, lakes, wetlands), water projects (reservoirs, irrigation channels, trunk sewers, etc.), water source protection zones, immigration control hydrological monitoring section distribution diagram (1 : 250,000); diagrams of soil type, land use, rainfall contour of the target area; maps of the different crop types and acreage (1 : 250,000); shallow and deep groundwater distribution, the overexploitation distribution, groundwater depth change map (1 : 250,000) were included in the baseline survey.

ET baseline data: ET baseline data were established in different zones, and transferred to land use maps at a scale of 1 : 250,000. Other statistics included: total land area, arable land and non-arable land, the Urban ET rural ET, water quotas, crop area corresponding to the ET distribution, amounts of fertilizer and pesticides used, the average precipitation over the period of record, typical precipitation and monthly rainfall of different frequencies.

Water consumption for production, life and ecology: This was a statistical summary of total water resources, the total surface and total groundwater resources, allocation of environmental water, groundwater extraction, total domestic water consumption, water consumption by industrial production, and eco-water consumption.

Agricultural groundwater extraction baseline: Shallow and deep groundwater extraction and utilization, groundwater overexploitation in target counties and districts, groundwater pollution, and mapping of shallow and deep aquifers and location of monitoring wells at 1 : 250,000.

Table 2.2 Monitoring and Evaluation Reporting Plan of the GEF Hai River Project

Report Name	Submitting Unit (Project Office)							Way of Submitting
	Central Project Office	IOC	Zhangweinan Sub-basin	Hebei	Beijing	Tianjin Water Res. and Envir. Protection	Tianjin Urban Construction Committee TUDEP	
Work plan of baseline survey	✓							Submit after the establishment of the monitoring and evaluation organization
Report of baseline survey	✓	✓	✓	✓	✓	✓	✓	Submit one time after the start of the project
Project monitoring and evaluation index system	✓							Submit after the baseline survey
Monitoring and evaluation implementation plan		✓	✓	✓	✓	✓	✓	Submit after the baseline survey
Annual report template	✓							Submit before the submission of the first-year monitoring and evaluation report
Annual report		✓	✓	✓	✓	✓	✓	Before July
Medium-term adjustment report	✓							Before the assessment of mid-course adjustments of the project
Monthly report	✓							The first 10 days of every month
Minutes	✓							Within 5 days after the meeting
Final report templates	✓							Submit before the preparation of the final report
Final report	✓	✓	✓	✓	✓	✓	✓	Submit one time after the completion of the project

Table 2.3 Sources of Data for the GEF Hai Baseline Survey, 2004

Location	Data Source
Basin-level	National data relevant to integrated water resources planning taken from the Water Resources Bulletin, the Environment Bulletin and the Regional Statistical Year book; ET baseline is based on 1×1 km^2 resolution remote sensing satellite images and suitable algorithms
Sub-basin level	Data for the Hai River Basin and Zhangweinan Canal Sub-basin in the national integrated water resources plan, and supplemented with data from key control points, cities and counties, and based on the Water Resources Bulletin, the Environment Bulletin and the Regional Statistical Year book; ET baseline as noted above
Tianjin	As above, but focused on Tianjin municipal area
Key counties or districts	Data are those collected for key control points for water resources and environment management and used in the national integrated water resources plan; ET baseline is based on the data of 30 m $\times$ 30 m resolution remote sensing satellite images

Wastewater discharge inventory for pollutants discharged to rivers: Industrial, urban, and rural sewage, and discharge amounts of pollutants into the river system. Data included: total wastewater discharged, average daily sewage discharge, the average daily COD emissions, the average daily NH_3-N emissions, total discharge of major industrial pollutants, urban pollutant emissions, rural wastewater, pollution assimilation capacity of receiving waters in the target areas for COD and NH_3-N and those receiving waters that are unable to assimilate these wastes.

Wastewater inventory for pollutants discharged to the Bohai Sea: Survey includes the actual monitoring data; the name of coordinate of the area into the Bohai Sea or the exit section and distribution maps, the statistical analysis of water quality by season under conditions of much, average and little rainfall; total nitrogen, total phosphorus, COD, and NH_3-N. The Tianjin Construction Committee TUDEP project is part of the baseline survey but was conducted separately by the technical units of the Tianjin Construction Committee TUDEP project. An independent report was used for the Dagu Canal sediment and sewage discharged to the canal mouth.

Chapter 3

CONCEPTUAL INNOVATIONS AND THREE TECHNOLOGIES

3.1 NEW CONCEPTS - IWEM

Water resources in China are divided into water quantity and water quality, and are managed separately in law and in practice. Nevertheless, quantity and quality of water are two indispensible aspects that must be managed in an integrated manner in order to resolve problems of water scarcity, pollution control and ecological needs for water. Therefore, the project has adopted the concept of "integrated water and environment management" (IWEM). This is parallel to the concept of IWRM that is now well understood worldwide but has proven difficult to implement at a practical level. IWEM, as used here, focuses on the practical management of water quantity and quality at basin, sub-basin, and lower administrative levels.

IWEM is the common mission of MWR and MEP, and requires joint planning, coordination and supervision of both ministries. This project provides a large-scale opportunity to address the systemic problems of un-coordinated management noted above in this monograph, and to develop, trial, and implement new mechanisms that lead to unified management of water resources at all levels of government. Specifically, in GEF Hai River Project, IWEM means:

Establishing unified management mechanisms at all levels.

Integrating water function zones and water environment function zones.

Utilizing the standard river coding technology.

Information sharing through the KM platform.

Improving water quantity and quality supervision.

Determining ecological flows and environmental capacity.

Stabilizing groundwater levels, and improving the water environment.

Establishing unified water permits and discharge permits approval and management system.

3.1.1　IWEM in Action

Implementing IWEM required horizontal cooperation amongst the Ministry of Water Resources, Ministry of Environmental Protection, Ministry of Agriculture and Ministry of Construction. Institutional coordination mechanisms have been described above. At the local level these institutional mechanisms were adopted and supervised by local government (e.g. at the county level) which guaranteed that effective coordination would be achieved amongst the various agencies under their jurisdiction. Joint planning for water and environment management for all aspects of the project led to each jurisdiction creating IWEM Plan with short, medium and long-term objectives. This was done for all demonstration counties and for Tianjin. At the basin level and for Zhangweinan Sub-basin Strategic Action Plan was created through joint planning mechanisms developed under the project.

Each IWEM Plan required ownership by the level of government for which the plan was intended and included a full implementation agenda that was formally accepted by the responsible government. Each plan established specific targets for implementation and for expected outcomes that would be realized after this project was completed. This was the first time in China that this type of integrated planning and implementation had been achieved in the water and environment sector at the large basin scale.

3.2　NEW TECHNOLOGIES IN GEF HAIHE PROJECT

3.2.1　Remote Sensing for ET Management

An integral part of this project was the development of control methodologies that would achieve water savings in the agricultural sector. This reflects the fact that agriculture is the single largest water user and is responsible for the overdraft of aquifers throughout much of the Hai River Basin. In addition to the usual water-efficiency methods applied to irrigation, this project recognized that the largest loss was from evapo-transpiration (ET). By controlling ET, especially the non-beneficial component, water balance calculations indicated that there could be very large savings in water loss; these savings would be reflected in reduced need for groundwater extraction. Water balance calculations indicated that, over time, control of ET could lead to stabilization of groundwater levels. Given the extreme water deficit in the Hai River Basin, ET monitoring and management was a core technology for the project.

Because of the large area of the basin, conventional land-based ET measurements from meteorological stations was not a solution; therefore, a remote ET sensing and monitoring system center was established at the Haihe River Water Resources Conservancy Commission (HWCC), and at a branch centre of remote ET sensing and monitoring management in the Beijing Water Authority. These centres applied commercially available algorithm to remote sensing data to determine the amount and geographical distribution of ET from spectral images.

ET Quota Management

Using a water balance approach, an ET quota was established for each county based on precipitation and groundwater availability. The ET amount is converted to a volume equivalent; the industrial and municipal requirements are deducted (fixed according to the "Five-Year Plan"). Some amount is allocated to environmental purposes according to environmental priorities of the county. The remaining volume is allocated to irrigated agriculture and is allocated downwards to farm units according to the irrigated area of each farm. This amount may be adjusted upwards or downwards depending on specific spatial or crop priorities of the county. During the project, county ET was remotely monitored and steps taken by county governments to reduce ET to meet the quota through variety of measures including change in crop structure, use of plastic mulch, increased use of waste straw on the soil, etc..

The next step was to link the ET allocation (as volume equivalent) to the permitted withdrawal from groundwater in agriculture. This was trialed in Quantao County where the water withdrawal permit was calculated according to the ET divided by the irrigation efficiency. Because this amount is less than farmers normally use, the government has invested in increasing irrigation efficiency so that farmers can meet the ET allocation without decreased production. In fact, farmers have been able to increase income with decreased amounts of water, by moving to higher value crops with their improved irrigation systems.

3.2.2 Knowledge Management (KM)

There are a series of technical innovations that deal with data, models and decision-support systems encompassing a wide variety of issues within the project. At the simplest level, KM is an integrated database and data management systems that was developed and distributed to all the project offices at all levels of the basin. At its most complicated, the KM system provided modeling and decision-support functions for all aspects of the IWEM development process. The KM system is integrated with the ET monitoring system discussed above. The KM system contains the following main functions:

Data sharing platform: An integrated database platform for water and environment knowledge sharing, and to realize cross-departmental, cross-regional, multi-level, and multi-systematic exchange and data sharing.

An SOA-based information exchange and application control platform: This allows integration of existing application resources and data sources, resolves integration and sharing problems of heterogeneous system resources, and achieves information exchange between heterogeneous data sources and business applications.

Hai River coding system: This is based on a dynamic segmentation technology patterned after the US STORET system in which the river network is divided into river reaches, each of which has a unique numeric identifier. This allowed the two different river coding systems used for (i) water function zones by MWR and (ii) water environment function zones by MEP, to be integrated without having to merge the two into a single system. River reaches are used for identifying all data that are associated with points or reaches of river (outfalls, water intakes, pollution loads, permits, etc.). These data are essential for developing basin-wide water allocation options, pollution control options, etc..

A basin-wide simulation system: Simulates water flows across the basin and from land to sea. It allows simulation of water conservation and pollution control, and considers precipitation, evaporation, surface and groundwater, water yield and water quality. This allows gaming with a multitude of options for integrated river basin management and from which optimal solutions can be chosen. This approach, based on water consumption in all sectors, is established with remote ET sensing and monitoring as management tool, water functional zones as the management unit, and water rights as a management core concept.

ET management system: This is described above.

General Structure: Used for the IWEM planning of key cities, counties and districts along the basin, KM system realizes information sharing amongst the MWR and MEP organization of those cities, counties and districts and for the demonstration projects. The general framework is shown in Figure 3.1.

KM Components: There are three levels in the KM system: basin level, province or municipal level, and county level. According to its content, the KM system at the basin level is composed of four parts: (i) data processing and transmission; (ii) KM basic platform (including operation environment, information exchange and application control

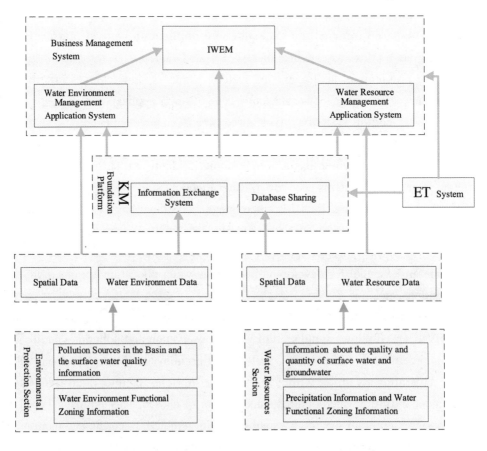

Figure 3.1 KM Framework for the GEF Hai River Project

Chapter 3 CONCEPTUAL INNOVATIONS AND THREE TECHNOLOGIES

platform, and database); (iii) business management system; (iv) standard norms and security systems. KM management tools at the county level include (i) data processing and transmission; (ii) basic platform at the county level (operation environment, and database); (iii) business management tools. The details are as shown in Figure 3.2.

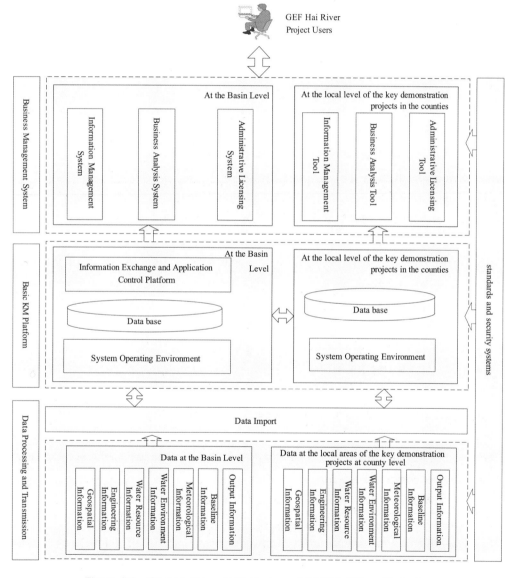

Figure 3.2 Composition of KM System of GEF Hai River Project

Data processing and transmission

KM at the basin level is a distributed structure, managed by HWCC and the Environment and Planning Institute of MEP. The two departments transmit and share their information through KM information sharing mechanism. KM at the provincial level is applied in Tianjin's Water Authority and the Environmental Protection Bureau. KM at the county level operates as a stand-alone system and is deployed at 16 key counties and demonstration projects. The information at different levels is shared through public internet, the water conservancy private network, and other media, as shown in Figure 3.3.

3.2.2.1 KM Platform and Key Technologies

To better realize data sharing, the followingtwo key technologies were developed for the KM platform:

(1) River coding technology based on dynamic segmentation

It was essential to build a cross-departmental bridge to different coding systems to exchange information such as water quantity and quality, pollution sources, function zones, etc. between MWR and MEP. Therefore, modeled on the USA's National Hydrological Database (NHD) river coding technology, a river coding system for Haihe River basin (HAIHE-NHD) was built by dynamic segmentation. As the basis for water information management, the coding system effectively reflects the correlation between reaches, catchment area, waters and related factors such as hydrological stations, sewage outfall, water quality monitoring stations, reservoirs, function zones, and retrieve and discharge outlet, etc.. The details are shown in Figure 3. 4.

In the GEF Hai River Project, dynamic-segmentation-based HAIHE-NHD contributes to all basin-related information sharing (including such point information as hydrology, water quality, sewage outfall, and linear information such as water function zones, and surface information such as land utilization, land property, agricultural nonpoint source and model calculation unit), and information connecting and exchange through river coding.

(2) KM information exchange and application control platform

The KM facility provides service for MWR and MEP at different levels at the basin, and requires cross-departmental, cross-regional, multi-level, and multi-systematic information exchange. In order to distribute database access pressure, and to ensure security of the database, a KM information exchange and application control platform

Chapter 3 CONCEPTUAL INNOVATIONS AND THREE TECHNOLOGIES

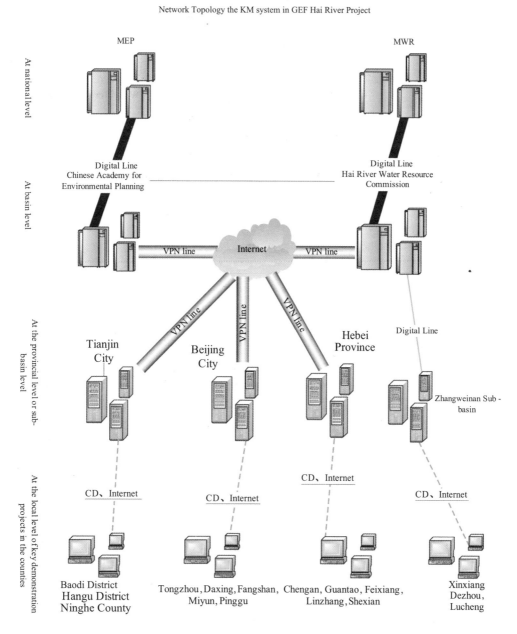

Figure 3.3 Distributed KM Deployment in GEF Hai River Project

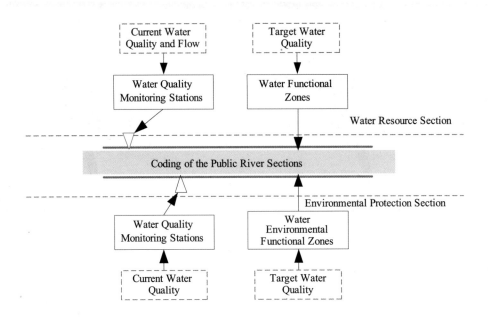

Figure 3.4 Relation Diagram of the HAIHE-NHD River Coding System

was developed to separate data and application systems. Through components and standard interface, this platform provides information sharing and data exchange support for ET system, dual model, and application system, and offers technical support to realize data sharing among many users and systems as shown in Figure 3.5.

Other key technologies included the water balance/circulation model, pollution estimation model, and the ET component.

Water circulation simulation system in the basin: The entire process of water circulation simulation from land to sea, from basin to region, from water conservation to pollution treatment, from precipitation to evaporation, from surface water to ground water, from water quantity to water quality, provides a multi-scenario analysis for IWEM. As the Hai River is water-scarce and seriously polluted a complete circulation model allows gaming with past and present conditions, options for water quantity and quality management, water levels control, groundwater extraction options, point and non-point source pollution, etc.. The objective is to establish sustainable options for surface and groundwater management and their consequences for the Bohai Sea. Figure 3.6 demonstrates how the model works.

Chapter 3 CONCEPTUAL INNOVATIONS AND THREE TECHNOLOGIES

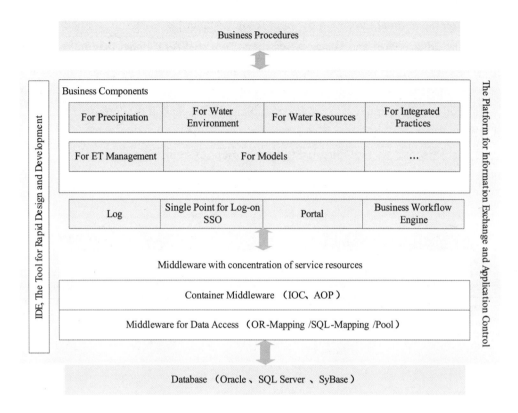

Figure 3.5 Structure of KM Information Exchange and Application Control Platform

This model is a complete water balance approach and consists of hydrological and water pollution components. An Ecological model (based on minimum flows) is a mathematical simulation of the precipitation-runoff water cycle in the studied area. The model includes precipitation, evaporation, surface-soil-ground water relationships, influence of natural conditions and human exploitation. It includes quantitative simulation of the distribution, forms of storage, and transfers amongst surface water, soil water and groundwater in the basin, as shown in Figure 3.7. The point and non-point source pollution component is illustrated in Figure 3.8.

The KM system uses two models to simulate the water cycle in the basin. The first is the Natural-Artificial Dualistic Water Resources & Environment Management Models (DWEMM) developed by Department of Water Resources of China Institute of Water Resources and Hydropower Research (IWHR). Based on natural water cycle, this model focuses on the optimal allocation of water resources and thus is mainly applied in basin level. The other is Soil and Water Assessment Tool (SWAT) model that addresses the influence of NPS pollution, water and soil erosion, land use and agricultural management on water quantity and water quality. SWAT can be applied to many different levels from basin level to sub-basin level to province/municipal levels and at the county level.

3.2.2.2 Business Management System

The objective of the "business" management system is to provide options for interventions in the water resources and environment planning process that can lead to

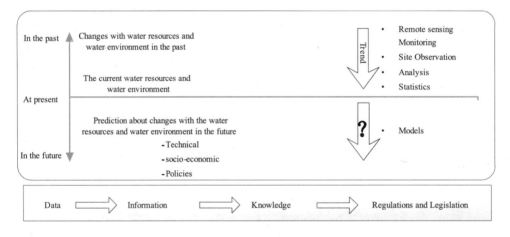

Figure 3.6 Process Diagram of How the Water Cycle Simulation Model Supports IWEM

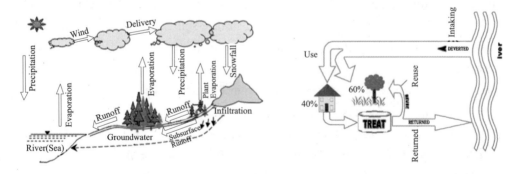

Figure 3.7 "Natural-Artificial" Water Cycle

Chapter 3 CONCEPTUAL INNOVATIONS AND THREE TECHNOLOGIES

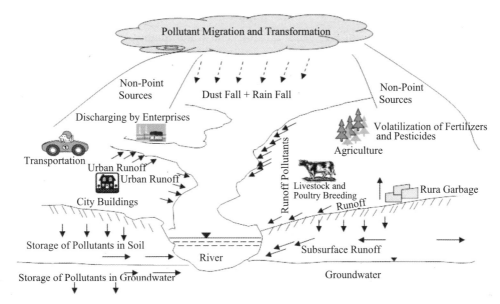

Figure 3.8 Point and Non-point Source Pollution Formation Process

effective outcomes. This business system is used at two scales – first, at the basin and sub-basin scale, and secondly, at the county and/or local scale.

(1) Basin Scale

On the 1 ∶ 250,000 GIS electronic map, integrated water and environment management zones, including water function zones and water environment function zones, are formed according to the criteria of the third province class water district. Through the monitoring of ET, water quantity and water quality, and the regulation of key areas, provincial boundary sections, and offshore sections, this system can portray the overall condition of water resources and environment, and predict their possible change under different management interventions. Then, plans of water resources allocation and pollutant control are proposed, and joint examination and management of water permits and waste discharge permit is implemented to achieve the reasonable allocation of water resources and control of pollutant emission. Figure 3.9 shows how the system works.

(2) County and District Level

On the 1 ∶ 250,000 GIS electronic map, integrated water and environment management zones, including water function zones and water environment function zones, are formed according to the geographical boundaries of counties and villages. Through the monitoring of ET, water quantity and water quality, and the regulation of key areas, and county boundary sections, this system can portray the overall conditions of the water resources and environment of the county or district level, and predict their possible change. Then, plans of water conservation and pollutant control are proposed, and joint examination and management of water permits and waste discharge permit is implemented through management of water permit, well drilling permit, sewage

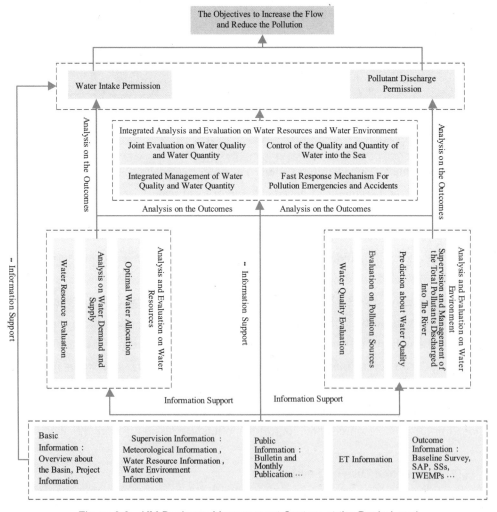

Figure 3.9　KM Business Management System at the Basin Level

emission permit, and WUA, to achieve the reasonable allocation of water resources and control of pollutant emission, and finally to realize water conservation and pollution control at the county level, as is shown in Figure 3.10.

3.2.2.3 Summary

To facilitate the KM program, the Department of Water Resources of MWR and Department of Pollution Prevention and Control of MEP have signed a data sharing agreement for the Haihe River water resources and environment, which guarantees sustainability of essential data for the KM system. The KM system assists in the development of "business" plans (planned interventions) for water resources and environmental development at basin and local scales and allows options to be evaluated relative to benefits within the basin and in the Bohai Sea. It complements the traditional sector approach to water management, promotes cooperation and coordination between the management departments at different levels, enhance IWEM and management efficiency, and sets a good example for China's basin management.

3.3 River Coding

3.3.1 The Stream Reach as the Fundamental Unit of a Surface Water Knowledge Base

Water resources planning is most efficiently done using a method in which all data (point and non-point information; discharges, intakes, river water quality, water biota, etc.) can be identified with a specific part of the river system. The method for doing this is the "river reach" system in which each "reach" of river is identified by its beginning and end point (often at tributary junctions or where there are significant changes in river characteristics). The entire river system is coded in which each reach is assigned a unique identifier. All other characteristics that contribute to the reach are stored according to the reach identifier. In this way, all attributes of the reach and of all other information that is related to water quantity and quality is stored within GIS system according to each river reach.

Because each of MWR and MEP have their own system of identifying river "zones" (water function zones and water environment function zones) the reach coding system allows the integration of these without having to merge one into the other. The reach coding connects different databases as shown in Figure 3.11. Point and line data are connected with river coding by event tables, and non-point data (such as land use) are connected with river coding by GIS overlay analysis, while dualistic and SWAT ecological models are also connected with data of land and soil use and sub-basin in similar ways.

3.3.2 Characteristics of the River Coding System

(i) Each reach has a unique address code. Any point, linear or non-point event can be related with only one code in the system.
(ii) The coding system includes upstream and downstream path search information, which can be programmed to achieve real time retrieval.
(iii) Connected with geographic information system, the system can reflect comprehensive and overall hydrological information and realize dynamic management.
(iv) The system is stable and expandable. Rearrangement of codes is unnecessary when the water network changes.

Figure 3.12 illustrates the reach system for the Hai River Basin.

3.4 ET Management

Evapotranspiration (ET) includes soil evaporation and plant transpiration. It is the link between hydrological process and ecological process in the basin, the link between energy and material balance, as well as the major mechanism of water consumption in agriculture and ecology. At the large scale water and energy exchange is affected by the local environment such as topography, terrain, geographical location and the underlying surface. The surface evapotranspiration of different surfaces can differ dramatically, therefore the understanding of spatial and temporal evapotranspiration structure can greatly enhance our understanding of the hydrological and ecological processes within the basin.

Because the Hai River Basin is extremely dry, the single largest loss of water is through evapotranspiration. This has two components – beneficial ET (water used by the plant for growth and metabolism), and non-beneficial ET (water that is evaporated from crops and agricultural surfaces but without performing any growth function,

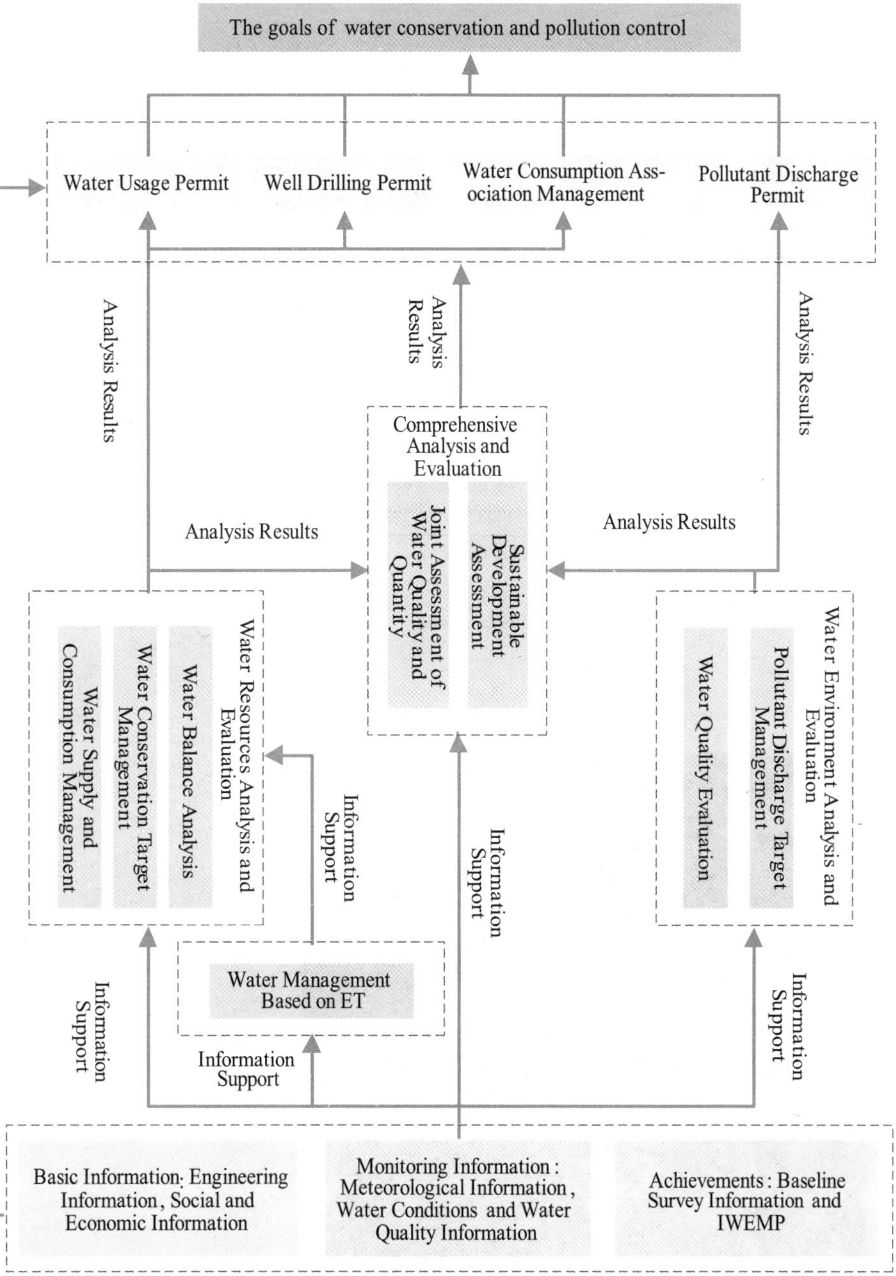

Chapter 3 CONCEPTUAL INNOVATIONS AND THREE TECHNOLOGIES

Figure 3.10 KM Business Management Tool in Key Counties and Demonstration Programs

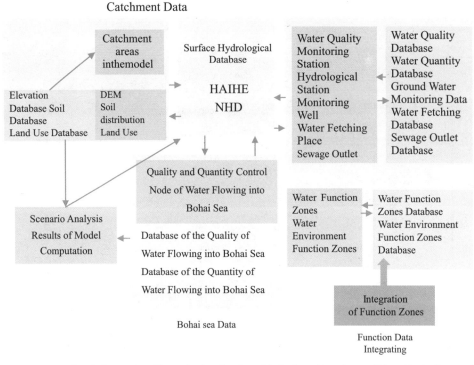

Figure 3.11 The Information Flows in the Surface Hydrology Database (River Path System)

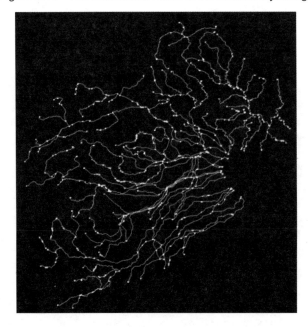

Figure 3.12 Reach System of Hai River Basin

such as excess irrigation water). The amount of water lost through non-beneficial ET is huge; management of non-beneficial ET would save vast amounts of water that are annually lost to the atmosphere and that should be left in the ground as groundwater and not pumped to the surface for irrigation and lost as non-beneficial ET.In this project, in water-scarce Hai River Basin, ET management has become an essential tool for re-evaluating the entire basin water balance and how to use water more efficiently, especially in agriculture. Therefore, ET management involves two essential steps – the first is the measurement of ET at basin and local scales; the second is the implementation of an ET management system in which farmers' access to groundwater for irrigation is restricted according to ET quotas that, in turn, limit the amount of water that is lost through non-beneficial ET.

To accomplish this task, three innovations have been developed in the field of ET management.

(i) A new model, "ET Watch",has been proposed based on the study of energy balance theory and evapotranspiration evaluation models, and of the topographic and climate features of Hai River Basin. This model uses studies of roughness's effects on ET in complex terrain, the time expansion of cloud contamination,data quality effects. The result has been a regional parameterization of models (Wu Bingfang, et al. 2008), and remote sensing estimates of ET has been developed to resolve water resources problems in the basin.

(ii) Using GIS, an ET monitoring and management system is established based on step #1 above.

(iii) An ET-based water resources management system is developed, including water balance analysis, ET quota, and allocation of water rights. This opens a new chapter in the application of ET to water resources.

3.4.1 Remote Sensing Estimate of ET

3.4.1.1 ETWatch

Due to the complexity of evapotranspiration, accurate estimation is affected by many uncertain factors, such as surface parameter retrieval accuracy, applicability of evapotranspiration models, time expansion, advection, local environment, etc..It has been necessary to make full use of the advantages of remote sensing in dynamic temporal and spatial monitoring, to resolve the conflict of instantaneous monitoring of remote sensing and continuous change of evapotranspiration. It also requires the suitable treatment and dealing of the relationship between theory and practice. Improved algorithms in theory without field verification are not necessarily useful in application.

To solve the above problems, the developer of ET management system put forward an operational remote sensing monitoring method – ETWatch, which combines "energy remainder" method and Penman-Monteith formula to calculate evapotranspiration. ETWatch uses different models according to the characteristics of remote sensing images and access to auxiliary data. When the resolution of the images is high, spatial variability is small and objects on land can be clearly differentiated, the SEBAL model and Landsat TM multi-band data are used to calculate daily evapotranspiration. In contrast, when the resolution is not that high, spatial variability is large, and mixed pixel images are in the majority, then SEBS model and MODIS multi-band data are used. Remote sensing models can often not get a clear image and miss some data because of weather conditions. Therefore, to obtain consecutive daily evapotranspiration data, the Penman-Monteith formula is introduced, which sets the evapotranspiration on sunny days as "key frame"; it constructs a time expansion model of surface impedance based on the surface impedance information in the key frames, and fills in the data loss caused by missing spectral images. It then uses the daily meteorological data to rebuild an evapotranspiration time series, and combines the temporal evapotranspiration change information in low resolution and the spatial evapotranspiration variance information in high resolution to build a high spatial and temporal resolution data set. ETWatch also provides evapotranspiration monitoring results at the basin level (1 km) and at the plot level (10 ~ 100 m) to meet the needs of water resources evaluation and agricultural water consumption management.

ET spatial and temporal data analysis:

With ETWatch model, the annual ETs of Hai River Basin in the period 2004-2009 are calculated. As is shown in Figure 3.13, the average annual ET from 2004 to 2009 was 467 mm while the average annual precipitation in the same period was 507 mm, which means a water "surplus" of about 40 mm that is available for all other uses (e.g. industrial and daily life water consumption, and water discharged to the Bohai Sea). The annual ETs from 2004 to 2009 are 480 mm, 452 mm, 444 mm, 457 mm, 473 mm and

Chapter 3 CONCEPTUAL INNOVATIONS AND THREE TECHNOLOGIES

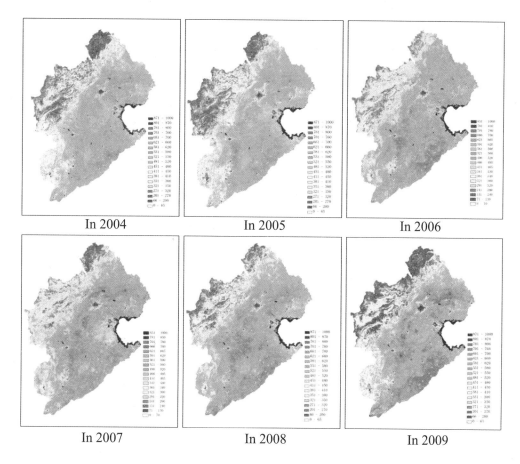

Figure 3.13 The Annual Evapotranspiration of Hai River Basin from 2004 to 2009

482 mm respectively. It was a normal flow year in 2004 and in 2008; it was a high flow year in 2009, while 2005, 2006 and 2007 were relatively dry years. Precipitation change ranges from -14% to 10%, while ET change ranges from -5% to 3%. The larger annual precipitation change than that of the annual ET change shows the storage function of soil and underground water in water utilization processes.

The integrated analysis of ET data in Hai River Basin and the monthly evapotranspiration (Figure 3.14) shows that the evapotranspiration change is generally similar with precipitation change. The average precipitation in Hai River Basin increases gradually from January to the first peak in May (80 mm). From January to May, Hai River Basin is in the dry season, with precipitation less than evapotranspiration, thus, in

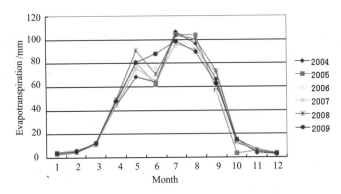

Figure 3.14 The Monthly Evapotranspiration in Hai River Basin from 2004 to 2009

water deficit. However, crops such as wheat in the basin are in their growth period and in desperate need of water. So, they can only depend on irrigation for their growth. With a slight decline in June, evapotranspiration again shows an upward trend and comes to its second peak in August, after which, ET decreases again. In June, the flood season comes, with precipitation increasing rapidly. In July, precipitation almost catches up with ET, and comes to its peak in August. From then until December, precipitation is more than ET, and the basin again moves to water surplus. However, due to the uneven distribution of water resources in the basin, some areas are still lacking in water even during water surplus period.

Different ETs in diversified underlying surfaces:

ET can be calculated according to specific land uses. For example, Miyun Reservoir ranks the highest with average ET above 700 mm. Paddy fields and reed plots follow with an average ET 600 mm. Plains, dry land, woodland, and shrub forest are similar in ET, with ET of around 500 mm. Grassland's ET is about 300 mm, and that of the inner city ranges from 100 mm to 200 mm, ranking the lowest among all measured land uses.

3.4.1.2 ET Data Generation and Management System

Based on the ETWatch model and ground observation data, an ET data generation and management system (Figure 3.15) was developed for HWCC. This system includes eight modules, namely: data acquisition, data preprocessing, ET data generation, ET model parameter calibration, ET analysis tools, search and retrieval, user management, and system management.

Data acquisition: The data used in this system includes remote sensing data,

Chapter 3 CONCEPTUAL INNOVATIONS AND THREE TECHNOLOGIES

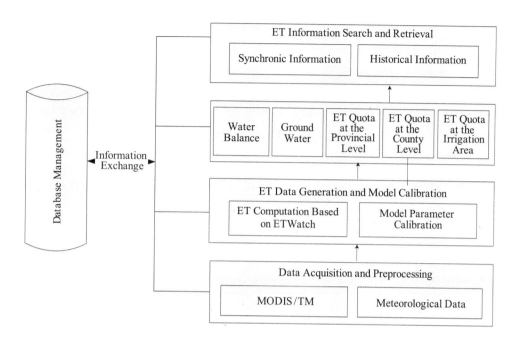

Figure 3.15 Structure of ET Data Generation and Management System

meteorological data, and other spatial data. Amongst three, the first two have relatively high access frequency, while other spatial data, such as DEM, slope gradient, slope exposure, regional boundaries, and land use, have a low update frequency, and don't need reprocessing. According to different spatial resolutions, remote sensing data are classified into low-resolution (1 km) MODIS 1B and high-resolution (30 m) Landsat TM5. MODIS 1B data can be downloaded from the NASA data distribution centre within one day as its time frequency. Landsat TM5 can be purchased from the satellite receiving station of the Chinese Academy of Sciences with 16 days of its time frequency. Meteorological data can be obtained from the Bureau of Meteorology with one day as its frequency. Tools are provided to import meteorological data into the system with target formats.

Data preprocessing: It is aimed at processing of remote sensing data and meteorological data. Obtained from original multi-band data, remote sensing data needs a series of preprocessing, such as geometric correction, radiation correction, atmospheric correction, and topographic correction, to generate the required surface parameters for the evapotranspiration model. The DN values of remote sensing images are finally transformed into surface parameter products with physical significance, such as the normalized difference vegetation index (NDVI), ALBEDO, LST, and cloud mask, etc.. Yet, since there exists a large difference between spatial resolutions of different sensors, radiometric calibration coefficients and methods to calculate surface parameters, MODIS and TM data preprocessing tools are embedded in the system. Meteorological data processing is to determine the space difference between different meteorological elements. This system provides five common difference algorithm and parameter settings.

ET data generation: ETWatch model is embedded according to required input and output standards to generate ET data. According to the direction of data flow, the process of data generation can be divided into four parts: i) the calculation of surface fluxes in fine weather through energy balance model and low-resolution land parameters such as surface albedo, surface radiation temperature, and surface roughness; ii) the calculation of low-resolution evapotranspiration products in consecutive days with time reconstruction algorithm; iii) the calculation of surface fluxes through SEBAL-class models and high-resolution surface parameters; iv) the forming of evapotranspiration products with high temporal and spatial resolution by means of data fusion algorithm.

ET model parameter calibration: As an important step in ET verification process, ET model calibration means to calibrate the intermediate calculation parameters with ground observation data, and to optimize the remote sensing parameterization scheme. The parameters are calibrated and optimized with the embedded PEST software in the system. With the optimized models, parameters are recalculated and compared with ground observation data until they meet the model's required accuracy.

ET applied analytical tools: These are tailored for ET. They include water balance analysis tools, groundwater tools, ET quote tools, ET quote tools at the county level and at the irrigation level.

Search and retrieval: Search and retrieval is ET information search tool developed using WebGIS. It retrieves ET real-time and historical information, which includes ET spatial distribution maps of different years and months, regional and land use types, and ET statistical information of crop types.

User management: This is the core of ET data generation and management system, and the connection point of data flow visit to all processing modules. It realizes user interface management of system database, and through data arrangement, analysis, and classification, builds a file-structured management system which is easy for users to understand.

System management: Since ET monitoring involves so many steps such as data acquisition, data preprocessing, and ET data processing and management, which relate to large data size and complex mathematical operation, the whole data processing chain is analyzed, and all the work is integrated into a top-down flow path, with each process linked to another. ET monitoring then realizes automation and streamlining.

3.4.1.3 ET-based Water Management Tools

The key to optimize and sustain water management is to reduce the current ET value, reduce exploitation of groundwater, and produce more surface water to achieve sustainable economic and social development in the basin. The control and abatement of ET needs ET allocation and management at different scales. Apart from a basin-scale ET management system, a top-down and hierarchical ET information management system is required. This requires five ET based water management tools.

ET-based water balance management tools: These ET based water balance analytical tools based on the SWAT model, and with the results from SWAT and related information, to analyze the water balance conditions in the third-class sub zones in different provinces from 2004 to 2008, and to develop scenario analysis modules to analyze the validity of strategic plans to achieve the target ET.

Specifically, ET based water balance tools are used to: i) obtain monthly and hydrological annual ET based water balance analysis of the basin; ii) obtain surface water inflow and outflow, groundwater inflow and outflow, and groundwater change, with SWAT model simulation and analysis; iii) analyze the water balance conditions of the third-class subzones in different provinces in 2003-2008; iv) achieve the integration of water balance tools and ET management system; v) provide situational simulation and data support for strategic research, Strategic Research Four in particular.

ET-based groundwater management tools: As an important component of ET management in the basin is to link the SWAT hydrological model with groundwater simulation. Therefore, we have developed groundwater management tools at the basin level based on RS ET and SWAT model, analyze the net consumption of groundwater in the plains areas of the basin in current conditions and multi-scenario situation. The objective is to identify the change in amount and distribution of groundwater usage under the various options for achieving the target ET.

These tools monitor and control groundwater in following ways: i) improve the groundwater module of the SWAT[1] model and obtain important basic data such as the net usage of shallow groundwater in the plains area of Haihe basin; ii) calculate the spatial and temporal distribution of groundwater net exploitation, soil water storage change and groundwater level changes with such basic data as precipitation, ET, and runoff volumes; iii) with the current net usage of groundwater and the spatial and temporal distribution analyses, and ET as a limit, identify those areas that exceed the allowable amount of groundwater usage; iv) track and simulate the net groundwater usage conditions under different land use and irrigation systems in the basin from 2006 to 2008, and work out the appropriate allowable amount of groundwater to stabilize the groundwater level; v) apply these results to identify water management options.

ET-based quota management tools at the provincial level: ET quota management at the provincial level takes ecological, social, and economic development into account, and tries to optimize the allocation of the provincial target ET to different county ETs. The allocation of the provincial ET quota is determined based on the following four aspects: i) the target ET worked out through water consumption balance; the target ET in different years is determined by the analysis of the historical water balance data, and the water control and groundwater excessive exploitation requirements in water resources planning; ii) ET-based water conservation potential; RS ET provides spatial water usage change. Through the analyses of water consumption of the same land use types and comparison of water consumptions of different land use types, the water conservation potential in the region is estimated, which rationalize the assessment of target ET; iii) the allocation of target ET; the adjustment range of the region is calculated through a comparison of target ET and actual ET; and with ecological, social, and economic development as well as water-saving potential in different counties taken into consideration, a non-linear goal programming model is established to realize optimized allocation of the regional adjustment range to the county level.

ET based water management tools at the county level: As an important component of RS ET monitoring application, ET based water management tools at the county level give technical support to water balance analyses at the county level, ET quota allocation, agricultural planting structure adjustment, and groundwater exploitation monitoring. Those tools are key to field ET control and include the following details: i) to realize information sharing with KM tools at the county level, and the statistical analysis of ET based RS data; ii) by ET quota allocation results and actual ET, to calculate water-saving potential of main crops, put forward water conservation measures and plans, and calculate county total ET; iii) to propose planting structure adjustment plan on the basis of ET quote allocation and water-saving potential analyses.

ET based water management tools at the irrigation area level: As an important component of RS ET monitoring applications, ET based water management tools at the irrigation area level bases itself on RS ET data, output data, soil moisture content, and crops distribution maps, and GIS platform, to realize visual management of water usage and consumption in different irrigation areas, and to give support to regional water availability, water permits, validate ET decreases, and groundwater levels.

The ET-based water management tools include the following: i) achieve platform sharing with KM tools at the county level, and map display and search based on GIS; ii) to estimate net irrigation water requirement according to the ET quota of main crops and precipitation; iii) to analyze the relationship between ET and the irrigation requirements of different precipitation levels, different soil types, and different crops with field SWAT model; and to provide definite ways to determine irrigation water quota of different crops, as well as the irrigation water quota of main crops in the region; iv) according to crops' different target ET and water-saving irrigation techniques, to adjust the frequency, time, and quantity of crop irrigation; to set up a reasonable irrigation system; and to reduce over-quota ET and water consumption by such three steps from criteria optimization to comparison of different plans, to determination of the best plan; v) to calculate water needs and crop duty graphs, and to propose water allocation plans in irrigation areas and water resources adjustment programs; vi) to set up an evaluation system of water usage management in irrigation areas, and to analyze statistically water usage management indicators and evaluate the effects of water irrigation and water-saving measures.

3.5 Development and Application of the SWAT Model and Dualistic Model

The SWAT (Soil and Water Assessment Tool) model (Arnold, etal.) is distributed process-based hydrological model for large and medium-sized basins. It can simulate a variety of hydrological processes, such as surface runoff, infiltration, lateral runoff, groundwater runoff, return flow, snowmelt runoff, soil temperature, soil moisture, evaporation, soil erosion, sediment transport, crop growth, loss of soil nutrients (nitrogen and phosphorus), water quality, and pesticides or insecticides, as well as the influences of such management measures as tillage, irrigation, fertilization, harvesting, and water allocation on the above processes (Figure 3.16). In this project, the SWAT model is mainly applied to NPS pollution simulation, spatial and temporal evapotranspiration simulation, and water balance analysis in Hai River Basin and Zhangweinan sub-basin.

3.5.1 SWAT Model in Zhangweinan Sub-basin

Enormously affected by human activities, Zhangweinan sub-basin has very complicated water usage relationships and suffers from incomplete data, which makes

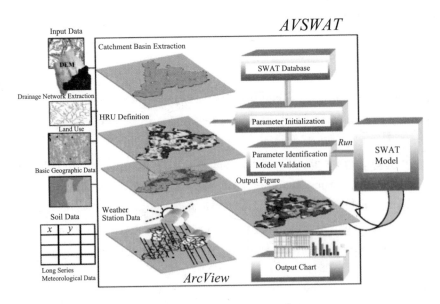

Figure 3.16 SWAT Model System

the hydrological and non-point source (NPS) pollution simulation very challenging. Much of the hydrology is artificially controlled with hundreds of sluices that control water flow according to demand within the basin. With spatial and attribute data, project researchers built a NPS pollution SWAT model system for the Zhangweinan sub-basin. Studies show that the main water pollution in Zhangweinan sub-basin is point source pollution, which is more severe in the sub-basins of Zhangweinan. Thus, meeting the standards required of point source pollution discharge is the top priority of water and environment protection and water quality improvement in the basin. NPS pollution is not common in the sub-basin. It mainly happens in the mountainous area of the basin, while in the plain area where winter wheat and summer maize are rotated, and surface runoff rarely occurs due to long periods of drought, loss caused by NPS pollution is much less, as shown in Figure 3.17.

The use of RS ET monitoring data can increase the validity of ET simulation, which, in turn, effectively improves the simulation accuracy of hydrological and NPS pollution processes. With 1 km resolution RS ET monitoring, the ET of the main farming areas in Zhangweinan sub-basin can be compared with those that have been simulated using the SWAT model. The SWAT model also allows exploration of farming measures which, with reasonable values for soil and crop growth parameters, makes the SWAT model more realistic in terms of actual conditions of the crop growth. Those research results provide experiences for reference for SWAT based ET simulation in north China, especially for regions with winter wheat and summer maize rotation.

3.5.2 SWAT Model of Hai River Basin

SWAT modeling has been a central technology for water balance and pollution forecasting at both large and small scales in this project. The following text describes the development and application of SWAT for the Hai River Basin.

3.5.2.1 Construction of Hai River Basin SWAT Model

On the basis of such underlying surface information as the digital elevation model, land use, and soil distribution, and after taking into account a variety of factors such as Hai River Basin's natural hydrological properties, the level three function zones, administrative divisions of provinces and cities, and distribution of hydrological stations and reservoirs, the Hai River Basin (including Zhangweinan sub-basin) is divided into

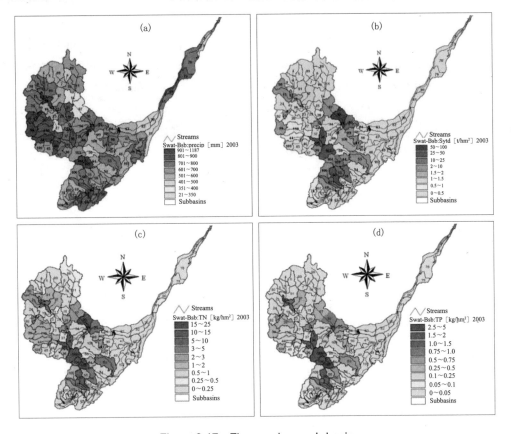

Figure 3.17 Zhangweinan sub-basin

(a) Precipitation and NPS Pollution; (b) Water and Soil Loss;
(c) The Total Nitrogen Loss; (d) The Total Phosphorus Loss

283 sub-basins, and 2,100 HRUs (Hydrologic Response Unit). This encompasses 17 estuaries, 48 major reservoirs and sluices, as shown in Figure 3.18.

3.5.2.2 Development of Graphical User Interface (GUI)

Based on Hai River Basin SWAT model, the GUI is developed for model adjustment, statistical counting and analysis of results, and scenario analyses. The GUI is shown in Figure 3.19.

3.5.2.3 Analysis of Water Cycle Elements

The elements of the water cycle and water balance in the basin are analyzed. This includes evaporation, precipitation, surface runoff, groundwater runoff, infiltration

Chapter 3 CONCEPTUAL INNOVATIONS AND THREE TECHNOLOGIES

capacity, etc.. Their relationships over a 12 month period are shown in Figure 3.20.

3.5.2.4 Analyses of Water Balance

The Hai River Basin SWAT model analyzes the monthly water balance conditions in Hai River Basin since 1995, including water inflow, water outflow, consumption, and profit or loss amount (Figure 3.21). Varied scenarios are set up to analyze the effects of water-saving measures (such as planting structure adjustment, change in irrigation methods, and preservation of soil moisture using surface mulch or other covering materials, on water consumption and water profit or loss.

3.5.3 Development and Application of Dualistic Model

3.5.3.1 Components of the Dualistic Model

The Hai River Dualistic Water Cycle Model, abbreviated as Dualistic Model was developed to reflect the dual nature of natural and socio-economic drivers of water management in this heavily populated basin. The model is composed of WEP (Water and Energy transfer Processes), ROWAS (Rules-based Objective-oriented Water Allocation Simulation), and DAMOS (Decision Analysis for Multi-Objective System), which couple together with each other.

- DAMOS is a macro level decision system that optimizes the allocation of water resources among different provinces and industries.
- ROWAS, through the modeling analysis of the allocation made by DAMOS,

Figure 3.18 Digital Distribution of River Network, Sub-basins and Reservoirs

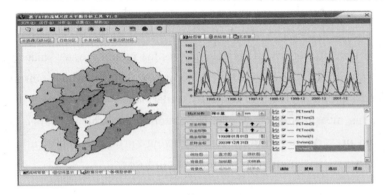

Figure 3.19 The Development of GUI Based on SWAT Model

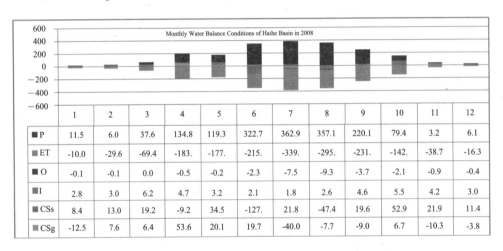

	1	2	3	4	5	6	7	8	9	10	11	12
■ P	11.5	6.0	37.6	134.8	119.3	322.7	362.9	357.1	220.1	79.4	3.2	6.1
■ ET	-10.0	-29.6	-69.4	-183.	-177.	-215.	-339.	-295.	-231.	-142.	-38.7	-16.3
■ O	-0.1	-0.1	0.0	-0.5	-0.2	-2.3	-7.5	-9.3	-3.7	-2.1	-0.9	-0.4
■ I	2.8	3.0	6.2	4.7	3.2	2.1	1.8	2.6	4.6	5.5	4.2	3.0
■ CSs	8.4	13.0	19.2	-9.2	34.5	-127.	21.8	-47.4	19.6	52.9	21.9	11.4
■ CSg	-12.5	7.6	6.4	53.6	20.1	19.7	-40.0	-7.7	-9.0	6.7	-10.3	-3.8

Figure 3.20 Monthly Water Balance Conditions of Hai River Basin in 2008 (billion m^3)

produces the water allocation process and scheduling for each management unit, and returns it to WEP after temporal and spatial distribution.

- WEP then gives its simulation feedback of water and environment evolution processes to DAMOS and ROWAS.

The models' coupling can be regarded as an information exchange and interaction process from macro information to middle level and to micro level. Figure 3.22 shows the three models' coupling relationships of research target, and results.

To achieve integrated water management in the basin and to take water resources, macro-economy, and ecology into account, WEP, ROWAS, and DAMOS divide Hai

Chapter 3 CONCEPTUAL INNOVATIONS AND THREE TECHNOLOGIES

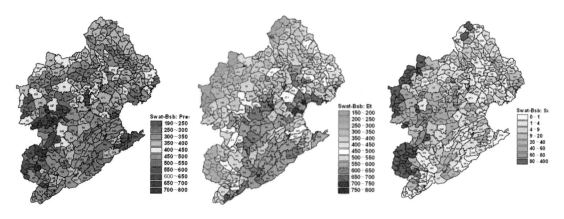

Figure 3.21 Spatial Distribution of Precipitation, Evapotranspiration, and Runoff of Hai River Basin in 2002

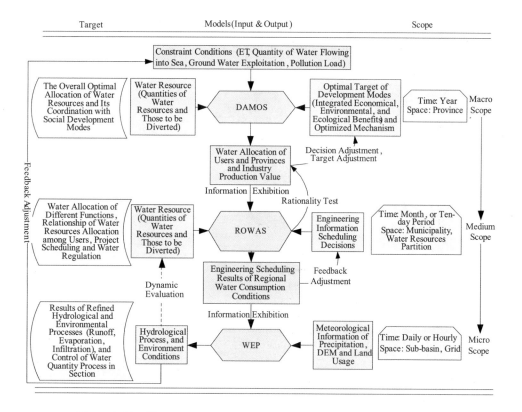

Figure 3.22 The Dualistic Model and the Coupling among Models

River Basin into 11,752 simulation units (contour belts in 3,067 sub-basin), 125 water resources planning/management units (15 sub-basins at 3rd level overlapped with 16 key counties), and 8 provincial administrative units respectively (Figure 3.23). Results show that the Dualistic Model is accurate in its simulation of ET, surface water quantity and quality process, groundwater flow field, and solute transport process for planning purposes and can provide scenarios analysis for water resources planning and management in Hai River Basin.

3.5.3.2 Summary of Results

Taking into account the basin hydrology and meteorology, SNWT, control of excessive groundwater exploitation, and control of water flow into the Bohai Sea, and using different base years, 17 scenarios were established for water resources management strategies using the Dualistic Model. Different indicators, such as GDP, grain yield, ET, into the sea water quantity and quality, total water consumption, pollutant discharge amount, decrease of excessive groundwater exploitation, are used to evaluate the 17 scenarios. The detailed results are shown in Table 3.1.

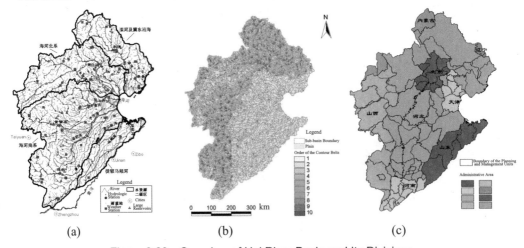

Figure 3.23 Overview of Hai River Basin and Its Divisions

(a) Distribution of Major River Systems, Hydrological Stations and Simulation Units; (b) Water Resources Planning/Management Units; (c) Provincial Administrative Units

Based on this output, seven controls (ET control, total amount control of pollutant emissions, control of excessive groundwater exploitation, and control of amount of water flowing into Bohai Sea, etc.) of water resources management strategies were carried out together by Beijing Institute of Water Resources and Hydropower Research (IWHR) and Tsinghua University in 2009. Results show that although the SNWT project can supply water to this basin and increase water environment capacity of the basin slightly, there will remain a need for strict control over water consumption and pollutant emissions in order to decrease over-exploitation of groundwater and increase water inflow to the Bohai Sea.

Table 3.1 The Results of Seven Control Indicators in Hai River Basin (S=Scenario)

Control Factors	Control Indicators	2005 Year as Base Year S1	2020 as Planning Year S10	2030 as Planning Year S11
Social and Economical Factor	Population(ten thousand)	11,970	13,558	14,031
	GDP (hundred million RMB)	23,717	104,587	137,260
	Grain Yield (ten thousand Ton)	4,415	6,519	8,020
Water Balance Factor	Precipitation (hundred million m^3)	1,599.7	1,599.7	1,599.7
Water Balance Factor	ET (hundred million m^3)	1,664.1	1,678.8	1,688.7
	Into Yellow River and Yangtze River Water (hundred million m^3)	41.3	117.3	157.0
	Outflow Water (hundred million m^3)	43.1	57.0	67.5
	Groundwater Storage (hundred million m^3)	-66.2	-18.9	0.5
Control of Water Usage	Surface Water Fetch (hundred million m^3)	135.6	230.2	260.6
	Surface Water Consumption (hundred million m^3)	102.6	147.5	162.1
	Groundwater Mining (hundred million m^3)	248.7	187.5	171.9
	Groundwater Consumption (hundred million m^3)	199.8	118.0	102.4
	National Water Consumption (hundred million m^3)	381.1	400.0	410.9
	Ecological Water Consumption (hundred million m^3)	3.5	17.1	20.9
Control of Water Pollutants	Chemical Oxygen Demand (ten thousand tons)	29.8	32.3	34.3
	NH$_3$-N (ten thousand tons)	1.45	1.55	1.64

Chapter 4

STRATEGIC STUDIES FOR WATER RESOURCES AND WATER ENVIRONMENT AND THE DEMONSTRATION PROJECTS

4.1 INTRODUCTION

With water ecology and environment deteriorating in the Bohai Sea, the goal of the Hai River Basin Integrated Water resources and Water Environment Management (IWEM) Project is to help alleviate the pollution in Bohai Sea by making integrated plans for water resource management and pollution control in the Hai River Basin. To this end, Strategic Studies(SS) and projects for water resource management and water pollution control were made in 8 major fields to provide guidance for making strategic actionplans(SAPs) for the Hai River Basin and the Zhangweinan sub-basin, as well as for integrated water resource and water environment management plans(IWEMPs) at the basin and regional levels. Additionally, demonstration projects were carried out in selected counties (cities, districts) in Beijing, Hebei Province, and Zhangweinan sub-basin. They focused on solving issues with water resource and water environment management, promoting the application and popularization of the real water-saving technologies, and strict groundwater management and pollution control. Demonstration studies played an important role in the implementation of IWEMPs in the project counties(cities, districts).

The 8 Strategic Studies(SS) are:

SS1 : Institutional reform, policies and regulations;

SS2 : Integrated water resource and water environment management in Bohai Sea;

SS3 : Restoration of the Hai River ecosystem;

SS4 : Efficient water use and water conservation;

SS5 : Groundwater management, water rightsand well permitting system;

SS6 : Wastewater reuse;

SS8 : Reasonable allocation of water resources in Beijing after the implementation of South-North Water Transfer.

Based on these eight strategic studies, IWEMPs were made for the project counties(cities, districts), and SAPs were made at the levels of Hai River Basin and Zhangweinan sub-basin.

4.2 STRATEGIC STUDY OF INSTITUTIONAL MECHANISMS, POLICIES AND REGULATIONS

4.2.1 Overall Objective

The objective was to establish an effective legal and institutional framework to strengthen the integrated management of water resources and water environment in the entire Hai River Basin, and at the same time toprovide guidance for developing effective legal and institutional framework in GEF Hai River project areas such as Hebei Province, Beijing, Zhángweinan sub-basin and the project counties (cities, districts) in Tianjin.

4.2.2 Suggestions for IWEM Policies in Hai River Basin

It is necessary to build an institutional system which contains water planning and management,water allocation management, water use management, water conservation and water environment management, public participation, and water administrative law enforcement.

The following institutional systems for water resource and water environment management need to be established or improved:
- new planningsystem
- regulatory system for the implementation and monitoring of planning process
- record and review system for regional planning
- water distribution system based on ET management
- system to ensure that the flow and water quality in the downstream sections in cross-boundary districts could reach standards
- ETquota management system
- groundwater exploitation control system based on ET
- water rights trading system

- water use auditing system
- charging system for use of environmental resources in the basin
- marine functional zoning and environmental protection planning system
- system for inspection and control of pollutant emissions
- system for approval of environmental cost in the basin
- upstream and downstream ecological compensation system
- emission right trading system
- system for sewage emission standards
- system for clean production and clean production management in key industries

The following existing systems need to be strengthened:

- water withdrawal permitting system
- total (pollution) load control
- water function zone management system
- system to evaluate the environmental impact of planning
- drinking water source protection system
- groundwater management system
- system for management of the inflow into the Bohai sea
- water pollutant emission control
- water resource monitoring system
- monitoring system for environmental quality and water pollutant emissions
- water conservation system
- water quota management system
- joint legal enforcement and inspection system
- water disputemediation
- water information publication
- the legal hearing system
- system for performance appraisal of water environmental protection responsibilities
- joint meeting system(co-management) for integrated management

Efforts, need to be made to strengthen management of water withdrawal, water consumption and water drainage to realize integrated management. An ET-based water rights system needs to be established in Hai River Basin, as a well as a system for total water intake control with ET management, an integrated water use system for the total regional water use control and quota management with water consumption control as the core part, a water drainage management system to control pollutant emissions into the river; to improve the drainage outlet setting agreement system in Hai River as well as the market mechanism for water resource development, utilization, and protection, and also for pollution prevention and control.

The water resource and water environment monitoring information needs to be further integrated for Hai River Basin. An overall plan needs to be made for the establishment of water resource and water environment monitoring network in Hai River Basin, and various information resources should be integrated by use of knowledge management (KM)system in Hai River Basin.

Policy management should be improved in Hai River Basin. More efforts will be made for the formulation and implementation of IWEMPs in the basin, to establish a coordination mechanism for the water resource protection and water pollution control in the basin, to clarify the water use rights and pollutant emission rights(water drainage obligations) of the upstream and downstream,to strengthen the integrated management of both water quantity and water quality of inter-provincial water bodies, and to establish a sound environmental and ecological protection compensation mechanism in the basin.

Supervision shall be enhanced. There needs to be a system to clarify the local government's water environmental protection responsibilities; the governments shall put more efforts in the regulation and legal enforcement;responsibilities shall be clarified for the local governments in the sections of water conservancy, environmental protection, agriculture and so on;monitoring of groundwater, surface water and pollution sources needs to be strengthened, as well as the monitoring of agricultural nonpoint source pollution; and there should be more public supervision.

4.2.3 Improving the Legal Framework

The law and regulation system for IWEM needs to be improved in the basin as soon as possible. The Groundwater Management Ordinance and the Drinking Water Source Protection Ordinance should be included in the annual legislative plan, and they should beformulated,promulgated and implemented as soon as possible. At the same time, those parts in the existing legal system that are too general need to be specified as soon as possible, in order to solve contradictions and conflicts between and within legal systems for resource and environmental protection.

The regulatory documents for IWEM in Hai River Basin need to be formulated and promulgated. The document, Views on the Implementation of Integrated Water Resources and Water Environment Management In Hai River Basin, needs to be formulated and promulgated by the State Council as soon as possible.

Departmental regulationsin favor of IWEM in Hai River Basin should be formulated, including the Integrated Water Resource and Water Environment Management Plan in Hai River Basin, the Water Rights Allocation Management Measures in Hai River Basin, the Regulations for ET Monitoring and Distribution in Hai River Basin, Regulations on Groundwater Development and Use management in Hai River Basin, Regulation on Water Resource and Water Environment Monitoring Management in Hai River Basin, Regulation on Management of Flow and Water Quality in Inter-provincial Sections in Hai River, Regulation on Zhangweixin River Estuary Management, and Regulation on Public Participation in IWEM in Hai River Basin.

4.2.4 Suggestions on Institutional Reform for IWEM in Hai River Basin

4.2.4.1 Propositions for Institutional Reform for IWEM at Basin Level

For the recent period(2006-2013): To establish the joint conference system at provincial and ministerial level for IWEM in Hai River Basin; to enhance the authority of the Hai River Water Resource Protection Bureau, and to improve the integrated management capacities of institutions in Hai River Basin.

For the mid-term(2014-2020): To set up coordination committee for IWEM in Hai River Basin; to improve the institutional setup in Hai River Basin, and to improve the integrated management of institutions in Hai River Basin.

For theLong-term (2021-2030): To straighten out the central system for resource and environment management and establish a management office in the basin; to set up the National Environment Monitoring Center, with a sub-center in Hai River Basin; to set up the Hai River Basin Integrated Management Committee.

4.2.4.2 Institutional Reform for IWEM at Sub-basin Level (Zhangweinan)

Under the auspices of the Ministry of Water Resources and Ministry of

Environmental Protection, led by the Hai Basin Conservation Commission, and based on the Zhangweinan Canal Administration Bureau, a management committee is to be established with the participation of water conservancy and environmental protection agencies at provincial,city, and county levels,as well as participation of representatives of water users. This management committee will be mainly responsible for the decision-making, implementation and supervision of the IWEM issues in Zhangweinan sub-basin, and will have aunified command over the water planning,flood and drought control, water distribution, comprehensive development and utilization, water environment security,water pollution control, etc. in Zhangweinan sub-basin.

4.2.4.3 Proposals for Institutional Reform for IWEM at Local Level

Systems will be established and improved for the integrated management of regional water affairs. According to the principle of *"Politics Separated From Business, Governments Separated from Administration, Clear Rights and Responsibilities, and Coordinated Operation"*, it is recommended that local Water Authorities be established as soon as possible, and water-related issues should be put under the full charge of these Water Authorities.

Leading teams for IWEM will be set up at the county(city, district) level, responsible for solving the major regional IWEM issues; offices will be set up under the charge of leading teams which should be established in the water conservancy department, mainly responsible for daily administration; a joint group of experts will be set up, mainly responsible for providing technical and advisory services.

Pilots for water user organization reforms will be carried out. Water Users Associations (WUAs) will be established or improved, and efforts will be made to set up the mechanism of "Community Driven Development"(CDD).

4.2.5 Implementation Plan for Reform

4.2.5.1 For the Recent Period (From 2006 to 2013)

The functions of all the relevant water organizations will be clarified. By going through the existing policies and regulations, revision plans will be made for the policies and regulations about water resource and water environment management in the basin, and drafts will be made.Joint conference systems will be set up at the provincial and ministerial level for IWEM in Hai River Basin together with an expert advisory committee. More authority will be given to Hai River Water Resource Protection Bureau with re-establishment of dual leadership of the Water Resources Protection Bureau by the Ministry of Water Resources and the Ministry of Environmental Protection, and clarification of its responsibilities for the supervision and management of flow and water quality in inter-provincial sections, and in implementing and supervising the IWEM in Hai River. Typical project areas will be selected for the demonstration of the joint approval system for water permits and discharge permits.

4.2.5.2 Mid-term Plan (From 2014 to 2020)

Legislation and legal Framework: The Drinking Water Source Protection Ordinance and Groundwater Management Ordinance will be developed as soon as possible, as well as the Views on the implementation of IWEM in Hai River Basin,Regulation on Management of Flow and Water Quality in Inter-provincial Sections in Hai River Basin, Regulation on Zhangweixin River Estuary Management, the Regulations on ET Monitoring and Distribution in Hai River Basin, etc.. Platforms will be built for water information (data) exchange and sharing in Hai River Basin based on the Knowledge Management (KM) system. Based on the Water Cooperation Declaration, there will be additional comprehensive planning and coordination mechanisms, joint law enforcement and inspection mechanism for IWEM, unified monitoring and unified information release mechanism, mechanism for public participation, and early warning and emergency response mechanisms forwater pollution incidents, in order to further improve IWEM in Hai River Basin.

Institutional Reform: A Coordination Committee will be set up for IWEM in Hai River Basin, withan secretariat and expert panel under its charge. The Hai River Water Resource Committee will have dual functionsas both the agency on behalf of the water conservancy administrative department of the State Council, as well as the executive body of the Coordination Committee. If any conflicts come up between the dual functions, they shall be submitted to the water conservancy administrative department under the State Council and the Cooperation committee for solution. Functions of the Hai River Water Resource Protection Bureau will be further expanded, with authorities over the water permitting approval, which gives it power forwater quantity management, and therefore there will be an unified management of water quantity and water quality. As the implementation and supervision agency for IWEM in Hai River Basin,it will be fully responsible for the water Resource Protection and water Pollution Control in the basin. An IWEM committee will also be set up in Zhangweinan sub-basin. The regional water management systems will be further improved and reformed.Pilots will be made for reform of water user organizations, and water users' association system will be further expanded.

4.2.5.3 Long-term Plan (from 2021 to 2030)

Policies and Regulations: There are two main areas of recommendations: Firstly, if the Ministry of a Water Resources and the Ministry of Environmental Protection are merged in the future, a Water Resource Security Act in the People's Republic of China should be formulated on the basis of the existing Water Law and the Water Pollution Control Law, with all content relating to water resource development, utilization,conservation, etc.in the existing two laws integrated into the new Law. The former two laws can be repealed. Secondly, if there are no significant changes with the existing institutional arrangements, it is recommended to develop coherent legal or regulatory documents. In addition, there should come out the Integrated Planning for Water Resource and Environment Management in Hai River Basin, Regulation on Water Rights Allocation in Hai River Basin, Regulation on Public Participation in IWEM in Hai River Basin, and the Regulation on Groundwater Development and Management in Hai River Basin.

Institutional Reform: The central system for resource and environment management needs to be more systematic with a management office established in the basin. The National Environment Monitoring Center is to be set up, with a sub-center in Hai River Basin. The Hai River Basin Management Committee will be established.

4.3 STRATEGIC STUDIES ON WATER RESOURCES

4.3.1 SS4: Efficient Water Use and Water Savings

SS4 is about water saving and efficient water use in the basin, based on the principle of water balance, and is designed to maintain good ecological environment and sustainable economic and social development. The study used an average of 2003 to 2005 as the baseline year, the year of 2010, and 2020 for target purposes, and three areas for demonstration purposes. SS4 is based on the principle of water balance, and the objective is to retain farmers' incomes yet reducing groundwater and maintaining appropriate inflow into the Bohai Sea.By using current ET from remote sensing monitoring and water use information in the basin, analysis and evaluation were made about the current quantity of water use,water use efficiency and water saving potential. Target ET was proposed for the basin and the three levels of project areas given the current water use in the basin and under the premise of sustainable development of the basin. Comprehensive measures were also proposed to achieve the target ET. The technical route of SS4 is shown in Figure4.1.

The conclusions of SS4 are as follows: Given the precipitation and diverted water in Hai River Basin, in order to realize sustainable development in the basin, namely, to maintaina certain inflow into the sea, to ensure that the groundwater level does not drop, the most effective way is to control ET in the basin and set a target ET in the basin. For a basin or region, given its water resources, its target ET is the water consumption that could meet the requirements of sustainable economic development and ecological development while sustaining a virtuous ecological circle at the same time.The definition of regional target ET contains three factors. The first is that it should be based on the basin or regional water resource conditions; the second factor is to maintain a virtuous ecological circle in the area; and the third factor is to realize sustainable social,economic and ecological development.

The target ET is determined by the method of "top-down, bottom-up, evaluation and adjustment".The "Top down" method is to have the basin-level water balance analysis, and then water allocation is made from the basin level to lower regional levels(including allocation of precipitation, inflow quantity, diverted water quantity, groundwater overexploitation, outflow, flow into the sea, etc.). The "bottom-up" way is that, given the allocation scheme,the target ET is calculated based on the regional ET under different water conditions. The method of "Evaluation and Adjustment" means that qualitative or quantitative assessments are made for the target ET in different scenarios to evaluate their rationality and to make final recommendations for regional ET.

In order to have a clear and complete description of the precipitation, ET, inflow into the sea, the hydrological cycle between the inflow and outflow of regional surface water and groundwater, as well as the complex hydraulic relationship among various water users in the whole basin and three levels of water resource areas, this study used

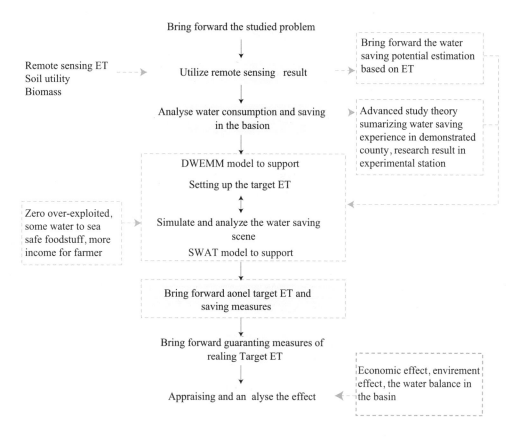

Figure 4.1　Technical Framework for the Strategic Study about Water Saving and Efficient Water use in Hai River Basin

the Dualistic Model for water resource and water environment management (described elsewhere in this monograph). With use of all these models, target ET for the basin and three levels of water resource areas were determined.With the analysis and calculation it was concluded that in order to achieve sustainable development in the entire basin (with no groundwater overexploitation and appropriate inflow into the sea), target ET in the basin for the years of 2010 and 2020 should be 523mm and 519mm respectively.

To achieve this objective, agricultural ET must be reduced. This means that, in order to ensure water security, food security and ecological security in the basin,and for the agricultural development with water-saving in Hai River Basin, the traditional management of water supply relative to water use should be changed to the management of water supply and water consumption; more effort should be put into comprehensive water conservation measures. Water conservation measures should be adapted to local conditions in order to improve utilization of water resources and increase crop water production rate. Some specific measures include:

- For the well irrigation areas that have groundwater overexploitation, water-saving irrigation measures and comprehensive water conservation measures can be taken tominimize the actual water consumption by crops, and to ensure that there will be no reduction in the crop production.
- In drainage irrigation areas where there is still potential for groundwater exploitation, besides the reformation of the irrigation areas, well irrigation and drainage irrigation should be combined.
- In areas where there is a shortage of surface water and shallow freshwater resources,shallow brackish water can be suitably exploited, and the irrigation technology with a mixture of fresh water and saline water should be promoted, in order to reduce the use of surface water and of deep confined water,and then to reduce ET while keeping stable crop yields.
- 53% of the basin is composed of mountainous areas. Given the scarce and disperse water resources in these areas, reduction of inefficient ET should be connected with soil and water conservation in order to increase farmers' income, and to have a stable condition for agricultural production.
- Crop planting structure should be adjusted to local conditions, and winter wheat plantings can be moderately reduced.
- Sprinkler irrigation, micro irrigation and other advanced irrigation techniques should be disseminated in the suburb areas of large and medium-sized cities, and areas with big plantations of vegetables, fruit and cash crops.
- In order to improve water conservation and efficient water use in Hai River Basin,ET management should be taken as the core concept for water management.

4.3.2 SS5: Sustainable Utilization of Groundwater, Water rights, and Water Withdrawal Permitting

The objectives for this study is to pilot a water rights and well drilling licensing management systemin Hai River Basin with ET management as the technical basis. The outcome is to reduce groundwater overexploitation to zero and achieve gradual recovery of the groundwater table.

The main tasks were:

- To complete the review and evaluation about the groundwater resources and its development and utilization in Hai River Basin.
- To propose a water rights management framework and a vision for total water withdrawal control in the basin, in order to improve ground water use efficiency, alleviate the contradiction between supply and demand, and commit to rational allocation of water resources in the basin.
- To complete an evaluation of the current situation with water rights and well permitting management in Hai River Basin.
- To propose a well permitting management system in demonstration areas to effectively control the excessive extraction of groundwater and to promote the sustainable water use in the basin and local areas.
- To make recommendations about using remote sensing monitored ET to manage groundwater water rights and well permitting to realize basin-wide sustainable groundwater use.

The activities of the study included the following major components:

- Investigations and assessments were made about groundwater development and utilization in the North China Plain areas of the Hai River Basin, the population supplied by groundwater, groundwater supply facilities, groundwater supply quantity, and overexploitation and pollution of groundwater development in the current year.
- Investigations of the status of groundwater rights and water permitting management including groundwater management practices, rules and regulations at the basin, provincial and municipal, county, and country levels. Good management practices were documented.
- Computational analysis on net Groundwater use: According to the current ET and target ET, and combined with the rainfall, irrigation by surface water, and groundwater recharge and losses, the net groundwater use was calculated for the 19 prefectures and cities in the plain area, which were then used to analyze and evaluate groundwater recharge and balance.

- The calculation of "three elements" of water rights(water withdrawal, water consumption, and water returned to the basin through runoff and percolation, etc.. Based on the recommended scenarios and the net groundwater use in the current year and in future target years, the "three elements" of water rights were calculated for the 19 prefectures and cities, and dynamic simulations of water level were made using the Modflow model in order to control groundwater exploitation and utilization.
- Groundwater licensing management framework:A licensing framework is proposed in terms of 10 aspects,including institutional mechanism,monitoring system, exploitation control by zones, target ET management, policy instruments, legislation, means, economic means, joint scheduling of surface water and groundwater, dissemination,training and research etc.. These are intended to guide groundwater management.

The study produced the following innovations:

- Based on the remote sensing monitoring ET, a new method was used to calculate net groundwater use, with the current ET and target ET taken into the calculation of groundwater net use and actual extraction.
- Description and calculation were made of the "three elements" of water (water withdrawal, water consumption, and water returned to the basin through runoff, seepage loss, etc., to control and manage groundwater development and utilization.
- Groundwater exploitation control measures were taken based on different zones in the plains areas.According to the current groundwater use and future requirements by the three elements of water rights, four types of areas were classified for groundwater development and utilization, which are the no-exploitation area,limited exploitation area, controlled exploitation area, and the conservation area. Relevant macro-countermeasures were proposed.

The following are the major conclusions of this study:

- There has been heavy exploitation of groundwater in Hai River Basin since the 1970s, with serious and continuing lowering of water tables and deterioration of the ecological environment deteriorated. In 2000, there were up to 59,600 km^2 with over-exploited shallow groundwater, an area of 56,100 km^2 with deep groundwater over-exploitation. Groundwater contamination is some 59,200 km^2. The study concludes that there are insufficient groundwater resources to support long-term sustainable economic and social

development in the basin under the present water management system.
- Remote sensing technology makes possible a new approach for ET monitoring and analysis in the basin, in which large-scale watershed and/or regional water balance including evaporation, become possible. Based on this,quantitative links can be established between remote sensing of ET and basin or regional groundwater net use that permits an assessment of groundwater use. There was an average groundwater net use of 6.21 billion cubic meters in Hai River Basin in each of the three years of 2003, 2004 and 2005.
- The target ET proposed for different baseline years could serve as effective guidelines for controlling excessive extraction of groundwater. With the target ET, the allowable groundwater withdrawal in Hai River Basin is 21.22 billion cubic meters, 20.6 billion cubic meters, and17.53 billion cubic meters for the current year, 2010 and 2020 respectively. With these extraction volumes, there was still overexploitation of groundwater in 2010, and a withdrawal and recharge balance of groundwater at the basin level is expected to be realized in 2020, with few slightlyover-exploited areas.
- Currently the water permitting management system is quite comprehensive and reasonable in Hai River Basin, but there is still room for improvement. The demonstration counties(districts) such as Guantao County in Hebei Province gained successful experiences in well drilling permit control and ET management, which can be disseminated to other areas based on their own characteristics.
- The three elements of water rights including allowable groundwater withdrawal, the maximum allowable consumption (ET), and water returned to the local water system, started new ways of groundwater management. By means of remote sensing technology, the difficulty in ET monitoring can be conquered and it can be used as the indicator to each of these three elements, in order to achieve the sustainable use of groundwater resources.
- Sustainable Use of groundwater is a dynamic process involving a relatively wide areas,and long periods of time. In order to achieve sustainable use of groundwater resources, the monitoring of groundwater information needs to be strengthened based on regional ET targets, and groundwater exploitation management should be compatible with different conditions in those areas.

This study provided the following guidance for the development of the two SAPs and the IWEMPs for the pilot counties:

Guidance for SAPs: The groundwater management framework can serve as reference for the future implementation of ET-based groundwater macro-management in the basin.

Guidance for the IWEMPs at county level: The groundwater management methods and experiences in typical counties can be disseminated to other areas and adapted to local conditions. The groundwater management system and methods can be used as reference for the county-level groundwater management.

4.3.3 SS6: Wastewater Recycling

With the serious water shortage and water pollution in Hai River Basin, wastewater recycling will not only improve water use efficiency, but also effectively reduce the amount of pollutants discharged into the rivers, which helps to ensure the realization of water quality objectives in water function zones.

In 2005, the total wastewater discharge in Hai River Basin was up to 4.485 billion tons, with 1,691,500 tons of COD emissions and 147,500 tons of ammonia emissions. By the end of 2005, there were 75 wastewater treatment plants in the Hai River Basin, with 24 in Beijing, 8 in Tianjin, 21 in Hebei Province, 6 in Shanxi Province, 8 in Henan Province, and 8 in Shandong Province. The total sewage treatment capacity was up to 8.185 million tons/day and 2.97 billion tons per year. For the 8 typical cities of Beijing, Tianjin, Shijiazhuang, Handan, Baoding, Datong City, Dezhou City and Xinxiang City, there are is total of 21 sewage treatment plants in the central urban areas, accounting for 28% of the total sewage treatment plants in the basin, with a designed treatment capacity of 4.14 million tons/ day, accounting for 50.6 percent of the treatment capacity in the entire basin.

According to the Classification Of Urban Wastewater Reclamation (GB/T18921—2002), urban wastewater reclamation that meet the standards can be used for industrial use, urban use and fluvial landscape. There were 383 million tons of recycled water in Hai River Basin in 2005, which was mainly used for industrial cooling and water recycling, municipal miscellaneous use, water supply for rivers and lakes, plant irrigation, etc., accounting for 19% of sewage treated in the basin.

There is both demand and potential for Wastewater Recycling. Recycled urban wastewater was mainly used for industries, river and lake landscape, plant irrigation, toilet flushing,car washing, etc.. According to the overall planning in the eight typical cities of Beijing, Tianjin, Shijiazhuang, Handan, Baoding, Datong City, Dezhou City and Xinxiang City, and taking into account the actual situation of these cities, as well as some constraints with wastewater reuse, there was a total demand of 883 million cubic meters for recycled wastewater in these typical cities, of which the industrial demand was the largest with 531 million cubic meters; the second largest demand was by river and lake landscapes for 203 million cubic meters. There will be a demand of 1.25 billion cubic meters of recycled waterby 2020, with an increase of 367 millioncubic meters from that of 2010 of which demand by industries will reach 665 million cubic meters, and 303 million cubic meters for river and lake landscapes.

The urban wastewater recycling potential depends on the processing scale of municipal sewage treatment plants and the production scale of recycling plants Urban sewage treatment plants in the central areas of the 8 typical cities, including Beijing, are to have a total of 6.695 million tons/ day and 8.125 million tons/ day in 2010 and 2020 respectively, and the production capacity of water recycling plants was and will be of 2.426 million tons/day and 3.263 million tons / day respectively, which should mainly meet the demand for recycled water.

Urban ET represents an amount of water that is mainly "wasted" as it is no longer available for urban water use. Reclaimed water, as part of the urban water resource allocation system, provides a stable water source for urban industries, for river and lake landscapes, for urban plantation and peri-urban agricultural irrigation, which leads to the increase of water consumption in the region. A Dilemma exists for local governments between the need for use of treated water locally, and the amount of treated water that should be passed downstream for other uses (including ecological use). Urban ET exacerbates the loss to downstream uses (Figure4.2).

Urban ET can be divided into the environmental ET,domestic ET and industrial ET. Based on the urban ET targets(Table 4.3.1) and the water resource allocation in the basin, urban water consumption can be controlled without expanding the scale of urban water use, and with more use of recycled water instead of fresh water extraction(surface water or groundwater). With wastewater recycling, the eight typical cities have almost the same

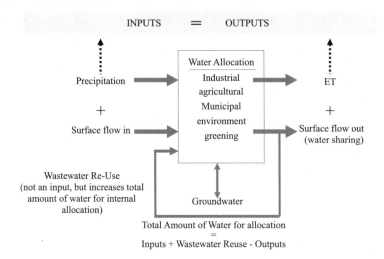

Figure 4.2 The Relation between Urban ET and Wastewater Reuse

outflow with sub-basins in the current year, which means that with urban target ET control, the urban water consumption and water use in the typical cities can meet the requirement of reasonable water resource allocationin the basin. Wastewater recycling plays a role in the reduction of urban fresh water use, and the freshwater saving can be used to increase the ecological water downstream or reduce the exploitation of groundwater, which is conductive for the sustainable development of ecological environment.

Table 4.1 ET Target in Typical Cities in Different Years   (unit: mm)

Year	Beijing	Tianjin	Shijiazhuang	Baoding	Handan	Datong	Xinxiang	Dezhou
2010	537	579	574	529	665	355	687	621
2020	560	626	585	594	529	368	614	608

In conclusion, wastewater recycling plays an important role in managing problems with insufficient water resources in the Hai River Basin. Treatment and recycling of wastewater needs to be fully integrated into the basin's water resource allocation system. Urban waste water recycling based on ET control can be demonstrated as an effective method to reduce both the extraction of fresh water and urban water consumption so that the urban area can meet the water resource allocation in the basin. However, wastewater recycling must be accompanied by measures to reduce urban water consumption. The key for urban waste water recycling is to establish and improve the wastewater recycling policies, and market mechanisms and investment environments. The key to reducing the consumption of urban water resources lies in urban industrial and domestic water savings,the reasonable control of the urban ecological environment, and reduction in loss of water through non-beneficial ET.

4.3.4 SS8: Reasonable Water Allocation in Beijing After the South-North Water Transfer

Beijing is faced with the problems of serious water shortage, groundwater overdraft, the use of strategic water reserves, lack of ecological and environmental water, inefficient water use, heavy water consumption, and large evaporation and transpiration. It is difficult to realize sustainable water use and integrated water resource management. The water supply pattern in Beijing will be significantly changed when water is transferred to Beijing by the South – North Water Transfer(SNWT) project. This will also be an opportunity for Beijing to optimize the allocation of water resources with multiple water sources and therefore to ease its water crisis.This study is designed, therefore, to produce efficient and sustainable water use programs, which target the sustainable use of surface water and groundwater,water consumption control, surface water quality control, etc..

Water allocation based on ET indicators is, in essence, water consumption management instead of water withdrawal management.The rational allocation of water resources based on ET indicators can be divided into three parts: analysis of regional or watershed ET targets (i.e.,water consumption indicators to meet the watershed or regional water balance),the target ET allocation and the validation of ET allocation plans with the use of water cycle simulation model.

A water allocation analysis model must first be established for allocation of ET in social, economic, ecological and environmental sectors. The water resource allocation model in Beijing consists of the following models: i) a comprehensive model to deal with the allocation of water resources for social, economic, environmental and ecological sections; ii) a water allocation simulation model to conduct detailed configuration based on the comprehensive analysis model; iii) the SWAT model to test and provide feedback for ET in the water allocation programs;and iv) a groundwater model for simulation and analysis for changes in groundwater tables.

Water Resources in Beijing: Water supply analysis was of surface water, groundwater, South to North transferred water, and reclaimed water, by considering the water supply objectives and quantity from all water sources. The total annual runoff in Beijing is 1.772 billion cubic meters, with 613 million cubic meters in the plain areas, and 1.159 billion cubic meters in mountainous areas. There is a total groundwater recharge of 228.684 million cubic meters in the plains areas of Beijing, and a total drainage of 256.386 million cubic meters. The exploitable groundwater in the plain areas is calculated to be 213.2million cubic meters.As described in the South to North Water Transfer Project Master Plan, the volume of transferred water to Beijing in 2010 was set to be1.052 billion cubic meters. According to the outcome of the Project Plan for the Middle Line of South – North Water Transfer, the net water transfer to Beijing in 2030 will be of 1.4 billion cubic meters.

Planning Scenarios in Beijing: Based on the current water resource development and utilization in Beijing, different scenarios were developed by taking into account specific scenarios as shown in Table 4.2.

Analysis of Water Allocation in Beijing: The total ET in Beijing is composed of natural ET and socio-economic ET. Socio-economic ET is the amount of water consumed by social and economic activities, such as agricultural irrigation, industrial water use, water for urban landscape, etc.; the other ET produced by the natural water cycle process is called natural ET. With ET targets set forvarious industries, and based on the regional water conditions, the water consumption of the various industries was determined. More focus was put on the industries with more effective water consumption.

4.4 STRATEGIC STUDIES ON WATER POLLUTION CONTROL

4.4.1 SS2: IWEM Strategy for the Bohai Sea

Since the late1970s, with the rapid economic development and urbanization in Hai River Basin and the Bohai Sea Rim area, large amounts of sewage discharge have created major environmental pressures for Bohai Sea and Bohai sea Bay. The marine environment has been deteriorating, with most of the estuaries and coastal areas now heavily polluted. There has been much more frequent incidents of red tide and other disasters. At the same time, with the rapid socio-economic development in Hai River

Table 4.2 Scenario Setting

Scenarios	Water from South-North Water Transfer	Pollution control	Groundwater over-exploitation	Inflow and outflow	Hydrologic period	Remarks
Scenario 1	Consistent with SS4	Consistent with SS2	Consistent with SS5	Consistent with SS4	25 years	Connected with SS2, SS4, and SS5, with integration of ET and water allocation in SS4
Scenario 2	50% water transfer, as planned.	Gradual reduction	In accordance with Dualistic Model	an average in 50 years	50 years	Connected with the results of 50-year series; reflecting the changes in water use and the corresponding groundwater in Beijing with a 50% decrease of water from South-North Water Transfer in 2010
Scenario 3	Water transfer as planned	fast reduction	no overexploitation by 2010	based on the needs of inflow into the Bohai Sea	25 years	Based on Scenario 1, there would be no excessive extraction of groundwater since 2010 with the water from transfer projects put in place. This scenario focuses on water conservation, which will facilitate the recovery of the groundwater table
Scenario 4	increase in the water transfer	gradual reduction	to realize no overexploitation gradually	an average in 25 years	25 years	Based on Scenario 1, increase in the water from South-North Water Transfer Project in 2010 will help to alleviate the ecological and environmental problems caused by water shortages in 2010; the impact of the midline runoff on the groundwater level will be discussed
Scenario 5	Delay of the water transfer	gradual reduction	consecutive dry periods	Consecutive dry periods	consecutive dry periods	There are descriptions about the water resources in Beijing with continuous drought periods ever since the year of 2000 and postponed water transfer from the Transfer Project

Table 4.3 ET Targets for Different Planned Scenarios   unit: 10,000,000 cubic meters

Scenario No.	Baseline year	Target ET	Natural target ET	Socio-economic target ET
S1	2010	87	64.04	22.96
	2020	88	64.04	23.96
	2030	88	64.04	23.96
S2	2010	99.27	73.46	25.81
	2020	101.29	73.46	27.83
	2030	104.78	73.46	31.32
S3	2010	97.36	72.46	24.90
	2020	97.36	72.46	24.90
	2030	100.78	72.46	28.32
S4	2010	101.21	72.46	28.75
	2020	98.59	72.46	26.13
	2030	102.01	72.46	29.55
S5	2010	83.17	64.04	19.13
	2020	88.21	64.04	24.17
	2030	88.21	64.04	24.17

Basin, there had been a sharp increase of water consumption which adds to water scarcity due to the small amount of rainfall in the basin. All these factors have caused a sharp decrease in runoff into the sea; in fact, there has been zero runoff into the sea in many areas. As a consequence, the salinity in the Bohai Sea keeps rising, creating a serious impact on marine ecology, particularly for spawning and nursery grounds for economic fish (which have low salinity requirements) with reductions in fish catch. This, together with over-fishing, has seriously hampered regional economic and social sustainable development. Control of pollutant loads into the sea, as well as increase of runoff into the sea is required to remediate the ecological environment of the Bohai Sea and to achieve sustainable socio-economic development.

This study concentrated on the two aspects of "total control" of pollutant load intothe Hai River – namely pollution reduction at source, and by dilution through increase in river flow. The work includes analysis of sea water quality targets, summary of pollution sources to the sea, establishment of a hydrodynamic model, the water quality response model,and a total control programming model.Analysis on the demand for increased flow is mainly the trend analysis of runoff into the sea in Hai

Table 4.4 Comparisons Amongst the Water Supply in Beijing in Different Planned Scenarios in 2010

Unit: 10^9 m^3

Data Type		Scenario 1			Scenario 2			Scenario 3			Scenario 4			Scenario 5		
		2010	2020	2030	2010	2020	2030	2010	2020	2030	2010	2020	2030	2010	2020	2030
Area scale	The entire city	39.38	41.93	43.05	43.33	47.61	51.42	42.10	43.71	47.61	42.69	44.92	48.97	34.61	42.32	42.47
	Urban area	24.81	28.43	33.10	26.42	31.15	34.68	25.96	29.80	33.58	23.11	29.63	33.51	20.12	23.49	25.50
	Rural area	14.57	13.50	9.95	16.91	16.46	16.73	16.14	13.92	14.03	19.58	15.29	15.46	14.49	18.83	16.97
Water source	Surface water	6.73	4.70	2.31	11.95	8.13	7.44	8.28	7.70	7.35	7.88	6.37	6.52	6.51	6.59	6.78
	Ground water	19.93	17.73	17.45	18.06	18.26	17.81	15.98	15.35	14.98	15.17	17.74	16.27	21.67	16.11	14.08
	Reclaimed water	7.44	8.99	9.28	8.07	10.66	12.20	7.30	10.11	11.29	7.64	10.27	12.18	6.43	9.13	11.12
	Water transfer	5.28	10.53	14.00	5.25	10.56	13.97	10.54	10.54	14.00	12.00	10.54	14.00	0	10.50	10.50

River Basin; the demand for fresh water runoff in Bohai Sea and the Bohai Bay was analyzed from the perspectives of water balance in the marine areas, the estuarine ecological water demand and the regression relationship between salinity and runoff. For the assessment of the effect of pollution reduction and increased flow, the first is to identify the various pollution abatement and flow increase measures, and then by using the Dualistic Model,increase of the flow was simulated and analyzing the effect on salinity by flow increases under different scenarios; there was also analysis on the impact by sewage emissions on the pollution in the marine areas.

Analysis on the structure of pollutant loads: There are land-based pollution and marine pollution in inshore waters. The land-based pollution is the major source for coastal water pollution in Bohai Bay. According to the years' monitoring data, land-sourced pollutants account for more than 80% of the total pollutants into the sea. The heavily polluted areas in Bohai Bay are mainly the estuaries and the coastal waters having direct sewage discharges.

Analysis on the Assimilative Capacity of Hai River Basin and the Design of Allocation Scheme: By using a linear programming model, and in order to meet the water quality standards in the environmental function zones of Bohai Bay, a fair, reasonable and cost-effective allocation scheme was developed by taking into account the factors of resources, population, etc..The results indicate that the current total emissions by different sources will exceed the water quality standards of the marine environmental function zones.

COD: The assimilative capacity of pollutant emissions into the Bohai Bay is up to 370,000 tons per year. However, with the current distribution of emission sources with different intensities, the actual capacity is less. With the current capacity of 300,000 tons of emissions into the sea, the basic functional zones tandards can be met. The allocable emission into the sea is between 140,000 and 270,000 tons per year according to the three different allocation schemes. Emission into Tuhai River needs to be reduced by more than 70% in these three allocation schemes, while there is a potential for an increase in other emission points.

Inorganic Nitrogen: According to the results from capacity estimation, it is acceptable to have a maximum of 22,000 tons of emission into the sea from 20 discharge points. The current emission of 33,000 tons is beyond the assimilative capacity, and results in inability to maintain compliance with the requirements in environmental function zones. Currently, the distribution of emission intensities causes some restrictions on the actual capacity. For example, the main load is concentrated in a few rivers like the Yongding River and Zhangweixin River. The distribution of Yongdingxin River in three different allocation schemes is only about 10% of the current emission. Therefore, the N load should be reduced by about 90% in order to meet the functional zone standards. However technically it is very difficult to accomplish this goal given the fact that the inorganic nitrogen loads are mainly non-point sources.

Inorganic Phosphorus: The estimate of assimilative capacity is a maximum of 2,100 tons into the sea from 20 discharge points. Currently, the distribution of emission intensities causes some restrictions on the actual capacity. In all the three allocation schemes, there needs to be a big reduction in Yongdingxin river, Zhangweixin River and Tuhai River. However it is very difficult to accomplish this goal given the fact that theinorganic Phosphorus loads are mainly non-point sources. However, it is easier than the control of inorganic N.

Quantitative analysis on the flushing of pollution loads during flood periods: The effect of the flood flushing of pollutant loads produces large volumes of pollutants with high concentration levels over a short period of time. This type of impact was simulated for the estuaries of Zhangweixin River and Hai River.For Zhangweixin River, the high concentration of load that might be flushed is mainly concentrated near the estuary; further upstream the impact by load flushing is weaker.Using a maximum concentration over 30 days as well as an average concentration that will exceed four different levels of water quality requirements, with the same estuary water storage, the higher the concentration, the bigger the impact will be by the discharges over the same time.

The overall impact on sea areas by load flushing shows a conspicuous trend of first increasing, then decreasing impact as the pollutant load in the estuary is transported seawards and then diluted as the impacted area increases in area. This means that there is initially a large area that exceeds quality standards but which progressively decreases in pollution concentration and area as the pollutant load is diluted and/or assimilated. For estuaries in the Hai River Basin, the scope of maximum concentration and average concentration exceeding Class Four water quality is similar to the load flushing in

Zhangweixin estuary. Taking into account the effects of maximum and average concentrations, it is acceptable to have around 5 days of emissions flushing load in Zhangweixin River, and around 3 days' emission in Hai River estuary.

Analysis on water balance and ecological demand of freshwater in marine areas: Based on the average freshwater flux data in the 20th century, and rainfall and evaporation data, estimations were made about the different water indicators in Bohai Sea from the perspective of water balance. Salinity levels are increasing; without taking into account the effect on salinity in Bohai Sea by exchanges between Bohai Sea and the Yellow Sea, it is necessary to increase the fresh water inflow into Bohai Sea to maintain the currentsalinity.

Analysis on the freshwater demand in estuaries in Bohai Bay: Estimation was made for water demand by the estuaries in Bohai Bay, based on the wetland ecological targets in the estuaries, the target water consumption by water cycle and ecological cycle, and the salinity balance target in estuarine habitats. The estimation results show that the total water demand for the estuarine ecosystem in Hai River Basin is approximately 7.87 billion cubic meters. Given that the estuarine ecosystem water demand only represents the basic water demand in the estuary, there is a now a serious shortage of runoff into the sea.

Analysis on the ecological replenishment scenario in Bohai Sea: Based on the results of the Dualistic Model simulation was made for salinity variances in the nine planning scenarios. Comparisons were made among the different scenarios in terms of salinity recovery, and propositions were made for water replenishment in Bohai Sea for the medium-term and long-term goals.

Suggestions on the IWEMP and SAP in Hai River Basin: Bohai Sea is faced with the requirement both of in-flow increases and pollution abatement. Yet increased flow may, paradoxically,cause a substantial increase of inflow of and non-point source pollutant loads into the sea, and thus result in a substantial increase in the frequency of red tide. However, no increase in flow will result in the disappearance of the spawning and nursery grounds (that have the requirement of low salinity). Therefore,more pollution abatement measures need to be taken to control pollution load to the sea while increasing the freshwater inflow into Bohai Sea and Bohai Bay.

Because reduction in runoff into the sea results in increased salinity in the Bohai

Bay, as well as changes in the ecological environment, it is necessary to specify the allocation rate of freshwater water to the sea within the water resource allocation scheme in the basin as a whole. Currently, all the freshwater resources in the Hai River Basin are used within the basin, especially considering that without taking into account the water diverted from the Yellow River in the total water quantity to the Bohai Sea, the runoff from the Hai River Basin is negative over the last decade. One of the tasks for the SAP and IWEMP in the Hai River Basin is to increase the inflow into Bohai Sea and restore part of the coastal habitat by using water from the South to North Water Transfer Project. It is necessary to have zero growth or a decrease in the water use in the Hai River Basin for the restoration of fresh water inflow into the Bohai Sea. In order to ensure the minimum ecological runoff, it is necessary to ensure the minimum runoff into the sea for the second quarter in dry years. As for the ecological freshwater inflow into the sea, it is recommended that within a positive estuary system, the sum of inflow into the sea and the precipitation in the marine areas should be greater than the average annual marine evaporation. Given that it is difficult to realize the above objectives in recent time, it is necessary to have them planned by different stages and periods in the basin plan.

Among the three main pollutants— inorganic nitrogen, inorganic phosphorus, and COD, it is relatively easy to control COD. It is difficult to control inorganic nitrogen and inorganic phosphorus, and especially to reduce inorganic nitrogen levels to comply with the short term requirements in the functional zones. Currently a more realistic approach is to gradually increase the denitrification and phosphorus elimination capacity of urban sewage treatment, and to promote the development of eco-agriculture and development of eco-technologies to reduce the use of chemical fertilizer so that there will be higher economic benefits attached to ecological agricultural products. More efforts need to be put to marine value-added stocking, and there should be strict management of the fishing industry, so that the annual catch can be maintained at a higher level, and the marine capacity of assimilating nitrogen and phosphorus shall be enhanced.

Most of the pollutants into the Bohai Sea come from gated seasonal rivers, mainly in the form of shock loading, which means that the wastewater is discharged suddenly before the arrival of a big flood, and therefore causes toxicity, hypoxia, and other hazards. As for this type of pollutant discharge, it is recommended that the storage capacity before the flood (mainly for storage of sewage in the river reservoir) should be kept at 40% of the designed capacity; when there is emission of stored industrial sewage in the case of flood, it should take no less than five days. The emission of stored domestic sewage into the sea for flood control shall take no less than three days; there should be more aeration facilities in the river to accelerate the degradation of organic pollutants and increase the dissolved oxygen level of water discharged into the sea.

It is a huge task to alleviate the ecological pressure on the environment, control water pollution, and improve the water environment in Bohai Bay. It is complicated to properly handle the relationship between economic development and environmental protection. Reasonable targets for water pollution control should be set based on the socio-economic condition; by rational utilization of water resources and effective control of water pollution, sustainable water use will play an important role in the sustainable socio-economic development of both land and sea areas.

Currently the pollution load into the sea and the sea water quality is stable, which is related with the reduction of flow into the sea. The increasing flow may result in an increase of pollutant load into the sea. Therefore, it is necessary to have the current water quality maintained or improved while increasing flow into the sea. There needs to be more pollution control efforts in the basin, particularly in the coastal areas.

4.4.2 Strategic Study 3: Water Ecological Restoration

Ecological Types of Rivers along Hai River Basin & Reaches and Wetlands that Need Special Protection: The sub-basins along Hai River Basin were classified into two categories. One group consists of sub-that are rich in sand and have developed large alluvial fans; the other group consist of sub-basins close to their mountain sources and have scattered tributaries and rapid currents. Each sub-basin is composed of rivers in the mountainous region and those on the plain. Among them, those on the plain can be further classified into four classes: rivers with clean water throughout the year, rivers with polluted water throughout the year, seasonally dry rivers or rivers that are dry in recent years, and long-term dry sand rivers.Dry rivers with poor-quality water that cannot satisfy the needs of its function zone should be given priority of restoration. Rivers and wetlands that are in good condition but need some special care should also be of top priority for protection.

Ecological Restoration Target and Water Consumption Needs under ET

Constraints: This report puts forward restoration targets for different waters in the mountain region and on the plains, including biological, water quality control, and ET targets.

Tree planting as the major way to convert areas with low and medium vegetation coverage into ones with high coverage in the mountains. Based on the HWCC Soil Erosion Control Plan that aims to complete 46% of the plan in recent term (around 2010), 36% in mid-term, and 16% in long-term, with additional ET consumption calculated as: 1.87 billion m^3 in 2010, 1.23 billion m^3 in 2020, and 1.52 billion m^3 in 2030.

Sand rivers can be gradually restored according to the available water supply from desert grasslands to wetlands, to wetlands with slow flows, to normal rivers. Their water quality should meet Class III. The recent ecological restoration target is to replace water with grass with an additional 150 mm ET, and restore rivers with grasses to rivers with water with additional 100 mm ET.

The ecological system in flowing rivers will be restored by means of communication channels, ecological sluice regulation, and costal morphology repair, etc.. The water quality should be $\leq$ Class 3. Rivers with water and no vegetation will be restored to vegetated state with additional 50 mm ET. The total ET increases about 86 million m^3 to achieve the ecological targets of sand rivers on the plain and rivers with water.

Wetlands will be controlled through the protection of water sources and their quality, and control of excessive deposition of biomass. Wetlands are planned to increase 294.3 km^2. Accordingly, ET will increase 50mm, and water consumption will increase 1.5 million m^3 according to remote ET evaluation.

A total 1.65 billion m^3 ET will be increased within the rivers in the mountains and wetlands, and on the plain.

Water Demand for the Ecological Restoration of Rivers & Wetlands on the Plain in Hai River Basin: The ecological water demand of the river system refers to the amount of water required to maintain its ecological structure and function. It includes the river's ecological base flow, evaporation loss and leakage loss. Taking the 1970s as reference and based on the historical record of 21 rivers, the HWCC put forward its Hai River Basin Water Resources Protection and Ecological Environment Restoration Plan. The ecological baseline flow is calculated in two ways: IHA (The Nature Conservation, 2006) is adopted in 17 rivers with relatively complete historical data, while the method of Tennant is adopted in 4 rivers with no or incomplete data to work out their ecological base flow. Their monthly ecological base flow allocations are also worked out through their monthly runoffs prior to human disturbance (mainly referring to the period before the controlling reservoirs were built).

Broadly speaking, the wetland ecological water demand refers to the amount of water needed to maintain the ecological system structure and normal ecological functioning. In the narrow sense, it refers to the amount of water needed for ecological consumption, namely its evapotranspiration consumption. The annual evapotranspiration loss of 12 key wetland areas included in the target planning of Hai River Basin is 612 million m^3, and the annual leakage loss is 844 million m^3, making the total water demand about 1.4563 billion m^3. And the annual water demand allocation is almost consistent with the annual allocation of evapotranspiration loss.

The water demand for the ecological restoration of 21 rivers and 12 wetlands in Hai River Basin is 5.6132 billion m^3 per year, of which the direct loss from rivers is 1.921 billion m^3 per year, and water loss into the sea is about 2.553 billion m^3 per year. Thus, an additional 2.9542 billion m^3 water is required to meet the demand for ecological restoration. In the current situation, only Dou River, Hai River main branches, and Majiahe River can meet the need; all other rivers are deficient in water to varying degrees. Based on the Dualist model simulation and the evaluation of ecological water demand, 13 rivers could meet the ecological demand in 2020, while the other 8 rivers including Dou River, Chaobai River, Beiyun River, Baigou River, Nanjuma River, Tang River, Zhang River, and Wei Canal cannot meet the demand. The water demand in 2030 is different from that in 2020. Yet, their water demand fulfillment conditions are the same.

Conclusions and Suggestion: Precipitation decreases caused by climate change, and water resources shortage and sewage emission increase caused by the rapid economic development are the major contributors to the degradation of the ecological systems in rivers and wetlands of Hai River Basin. Based on the available water supply and ET constraints, this study develops reasonable step-by-step restoration targets and models for different ecological types, and works out the total water demand to restore the ecology to the conditions in 1970s. Management mechanisms include:

- Reclaimed water quality standard will be fully promoted in the whole water body on the plain.
- Polluted source water treatment will be strengthened and water into the rivers

should be guaranteed to meet Class 3 level.
- Environmental flow managing mechanisms will be established using the system of sluices in the river system.
- Ecological evaluation mechanism and integrated management system will be established based on the rivers' ecological categories and ecological survey.
- Multi-disciplinary experts in ecology will be encouraged to join the design and implementation of irrigation projects.
- Water conservancy ecological effects monitoring and post-evaluation system will be implemented.

4.4.3 Strategic Study 7: Control of Pollution Sources

Partitioned Area Control Strategy: Based on the pollution characteristics of different administration areas in the basin, economic development differences, the key elements influencing pollution, and other 7 factors, Hai River Basin is divided into 4 different pollution control areas, as is shown in Figure 4.3. Different countermeasures are put forward for those areas.

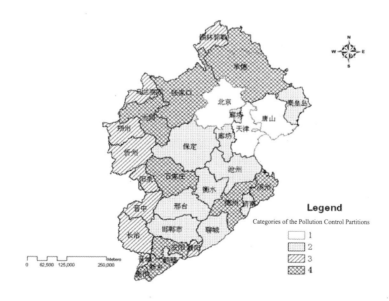

Figure 4.3 Pollution Control Partition of Hai River Basin into four categories

PC1 Class Area: refers to developed areas such as Beijing, Tianjin, and Tangshan, where technical support, capital investment, and enforcement of control take an upper hand than other areas. Domestic pollution control in urban areas becomes the focus.

PC2 Class Area: Refers to medium developed areas such as the southeast of Hebei Province, Jinan and Liaocheng in Shandong Province, and Hebi in Henan Province. In those lightly polluted areas, industrial pollution is the top contributor. Therefore effort should be focused on control industrial pollution. Domestic pollution stands out in a few towns and cities, and control efforts should be adjusted accordingly.

PC3 Class Area: Refers to underdeveloped areas such as Shanxi Province and Inner Mongolia Autonomous Region. In those lightly polluted areas, industrial development is constrained, and domestic pollution is dominant. Thus, the urban sewage pollution treatment plants should improve their capability and skills, and maximize the limited funding to the utmost for control of domestic pollutant emissions.

PC4 Class Area: Refers to moderately developing areas which are in the speeding process of urbanization and industrialization and are becoming industry-oriented, such as the northwest of Hebei Province, most areas in Shandong Province and Henan Province. They are now in a period that exerts excessive pressure on the local ecological environment. Industry-dominated, they should consider the transformation of their economic development.

Point Source Pollution Control Targets and Measures: Pollution control targets in 2010, 2015, 2020 and 2030 and correspondent measures are made respectively. Control measures of industrial pollution include (a)enforcement of sewage emission standards; (b)raising emission standards;(c) forced closure of illegal firms; (d) industrial restructuring; (e) centralized sewage treatment.

- Urban domestic pollution control measures include: (a) improvement of sewage treatment rate; (b) improvement of sewage treatment plants' operating load rate; (c) improvement of sewage effluent concentration; (d) improvement of the reuse rate of reclaimed water.In 2010, cities are required to further enhance the fulfillment rate of corporate pollution emission standards, and their total pollutant emissions should reach the 11th "Five-Year" Water Pollutants Reduction Responsibility Documents signed between local governments and MEP. The total pollutant emission in Hai River Basin should be within the emission scope required in the 11th "Five-Year"

Water Pollution Prevention and Treatment Plan (Table 4.5).

- To meet the emission reduction target, PC1 Class areas should increase their fulfillment rate by 5%. PC3 Class areas need a 15%~20% increase, and PC3 Class areas need a 20%~30% increase, while PC4 Class areas should increase the rate by 30%~40%. The key to the urban sewage control is to keep the normal operation of sewage treatment plants and guarantee their operating load rate above 85%. The sewage effluent concentration of the treatment plants in PC3 Class areas should meet Level 1-B, while the concentration of the plants in other areas should meet Level 1-A.

Table 4.5 Pollution Control Target of the Four Areas to Achieve the Hai River 11th "Five-Year" Plan in 2010

Area	Increasing the fulfillment rate by/%	Domestic sewage effluent concentration	Increase the plants' operating rate by/%
PC1	5	Level 1-A	80
PC2	15-20	Level 1-A	75
PC3	20-30	Level 1-B	75
PC4	30-40	Level 1-A	75

In 2015, PC1 areas cancel industry emission standards and all enterprises should practice the local integrative sewage emission standards. Specifically, the effluent concentration of COD and ammonia nitrogen should be 60mg/L and 10mg/L respectively in Beijing and Shandong, and 60mg/L and 8mg/L respectively in Tianjin. The standards in other cities and provinces are 100mg/L and 15mg/L (80mg/L and 15mg/L respectively in papermaking and food industry). All enterprises are required to meet the discharge standards on the premise that they should raise the sewage emission standards.

All the heavy polluting small firms with small economic scales and excessive pollutant emissions should be closed. In 2015, all heavy polluting firms without industrial scale should be closed in PC1, PC2 and PC4 areas. Only non-pillar industrial firms with small scales and heavy pollution should be closed in PC3 areas. Such local dominant industrial firms with small scales as papermaking and foods can be optionally closed. The operating load rate of the urban sewage treatment plants should be above 80%, and above 85% in relatively developed areas. The effluent concentration of COD, BOD5, total nitrogen, ammonia nitrogen and total phosphorus should reach Level 1-A.

The establishment of sewage treatment facilities at the county level should be given attention. In Hai River Basin there are 232 counties, cities or districts, among which 37 built 42 sewage treatment plants in 2007. Counties, cities or districts with daily domestic swage discharge above 10,000 m^3 have to build sewage treatment facilities, and the amount with treatment facilities should reach 75%.

All areas are required to reach the standard by 2020 on the premise that the sewage emission standards should be raised. The effluent concentrations of COD and ammonia nitrogen should reach 50mg/L and 8mg/L respectively in Beijing, Tianjin, and Shandong Province, and 80mg/L and 10mg/L respectively in other provinces.

In PC1, PC2 and PC4 areas, firms with the aggregate output less than RMB 10,000 Yuan in mining, chemical engineering, pharmaceutical manufacturing, metallurgy, machinery, and petroleum coking should be shut down. Firms with the aggregate output less than RMB 50 million Yuan in textile, leather manufacturing, papermaking, foods, and wood processing should be shut down. In PC3 areas, all firms with small scale in polluting industries should be shut down. Urban sewage treatment facilities' operating load rate should reach at least 85%, and the average processing rate should be above 80%. Eight additional cities' sewage treatment facilities' capacity reach above 10,000 tons per year.

The reuse rate of the reclaimed water should reach above 30% in Beijing, Tianjin and Tangshan, and above 20% in other areas.

In 2030, water environmental capacity target should be reached. Higher requirements will be met in urban domestic sewage reduction in Hai River Basin. All industries realize the integrated sewage emission standards, and the concentrations of COD and ammonia nitrogen shall be 50 mg/L and 8mg/L. All firms with small scales or with aggregate output less than RMB 100 million Yuan should be shut down. High-tech industries with vigorous economic returns and low pollutant emission will gain more support; the development of high polluting industries will be constrained, and efforts will be made to practice industrial restructuring and optimize industrial structure. With those efforts, the industry can finally develop in a healthy way.

- Urban sewage treatment should operate in full operating load rate in Beijing, Tianjin and Tangshan and above 90% in other areas. The average urban sewage treatment rate in the basin should be above 90%, and 99% in Beijing, Tangshan, and Qinghuangdao. New sewage treatment facilities are newly established in some cities, and eight cities are newly equipped with treatment facilities with an over ten thousand tons per year capacity. On the premise that the sewage treatment capacity is improved, they should continue enhancing the recycling rate of reclaimed water. The rate is set at 40% in Beijing, Tianjin and Tangshan and at 30% in other areas.

Pollutant Reduction Control Plan: According to the above measures and plans, the pollutant control targets and pollutant reduction rates in different periods are made, as is shown in Table 4.6. By this, the 11th "Five-Year" Plan of Hai River Basin can be reached in 2010, and the water environment capacity target in Hai River Basin can be achieved by 2030. COD emission and ammonia nitrogen emission should be within 107.02×10^4 tons, and 12.61×10^4 tons respectively in 2010, which decrease by 13.7% and 14.3% compared with those in 2005.

2015: COD emission and ammonia nitrogen emission decrease by 60% and 55% respectively compared with those in 2005.

2020: COD emission and ammonia nitrogen emission should decrease by 74% and 79% compared with those in 2005.

2030: COD emission and ammonia nitrogen emission should decrease by 92% and 98% compared with those in 2005. Water pollutant emissions in the basin should be below the water environment capacity. The capacities of COD and ammonia nitrogen in the basin are calculated to be 11.88×10^4 tons and 0.442×10^4 tons respectively.

Table 4.6 The Control Target and Reduction Rate of Pollutants in Hai River Basin

Control Targets	Targets of the 11th "Five-Year" Plan	2010	2015	2020	2030	Water Environment Capacity Targets
COD Control Target/10^4 tons	123.7	107.02	49.58	32.81	10.18	11.88
Reduction Rate Compared with Those in 2005/%	14.2	13.7	60	74	92	90.4
Ammonia Nitrogen Control Targets/10^4 tons	14.71	12.61	6.62	3.06	0.365	0.442
Reduction Rate Compared with Those in 2005/%	14.2	14.3	55	79	98	97.0

Pollution Control Strategies of Serious Polluted Industries: With many papermaking manufacturers, Hai River Basin is one of the largest papermaking regions in China. In the 2007 nationwide output rank of papermaking and paperboard, Shandong, Henan and Hebei Provinces in the basin remain the first, third and sixth respectively. In 2007, their gross industrial output value of papermaking, sewage emission, and emissions of COD and ammonia nitrogen account for 2.7%, 30.9% and 16.8% of the total of the Hai River Basin.

Under the three standard emission scenarios, COD emission caused by papermaking will be reduced by 70.8%, 76.7% and 85.4% respectively. In Scenario One (100mg/L), the percentage of COD emission caused by papermaking in most areas along the basin among the total emission will decrease to below 30%. In Scenario Two (80mg/L) and Scenario Three (50mg/L with special emission restriction), the COD contribution caused by papermaking in the basin will decrease to below 20%. Under the small-scale enterprise closure plan, 283 unqualified enterprises are to be shut down and COD emission reduces about 73,000 tons. The reduction rate of COD in papermaking will reach 28.4%, declining from 40.3% of the total COD emission in 2007 to 28.8% of the total. However, the papermaking industry in the basin is scarcely influenced with only an 8.4% decrease, from 2.7% of the gross industrial output to 2.5%.

Regional Conflict Management Strategies in Zhangweinan Basin: Regional conflict in Zhangweinan Basin is featured by the conflict between water quality and water quantity, conflict between upstream and downstream, and conflict between the left bank and the right bank. The Dualistic model simulates water pollutant concentration under designed water quantity and reduction plans. Designed water quantity includes three situations under underground water utilization, seawater utilization and SNWT. Pollutant reduction plan adopts rates at 10% and 20%. Simulation results show that the average pollutant concentrations are still far from the standard. In particular, the concentration rates in Longwang Miao and Qimen are five times or more higher than the standard.

Based on the cross-regional conflict coordinating model with satisfactory targets, Shanxi, Henan, Hebei, and Shangdong Provinces are taken as the conflict subjects to adjust and optimize their water consumption and pollutant control by taking their water quantity (ecological flow) and water quality (COD's and NH_3-N's concentration targets) into account. Results show that if the flow and pollutant concentration need to reach the targets for trans-boundary locations, then the allowable aggregate water extracted in the basin should decreaseby 55.13% compared with the current amount; the allowable emission of point source COD and point source NH_3-N should decrease by 87.53% and 91.48% respectively compared with current emissions. The reduction rates are a bit less in Shanxi and Shandong Provinces. The reduction rates of COD and NH_3-N should be 91.5% and 93% respectively in Henan Province with both above 76% in other provinces.

To manage interjurisdictional water pollution two economic models are suggested. Under the dual target control on water quality and total pollutants in the vertical compensation model, Hebei, Henan, Shandong and Shanxi Provinces should pay RMB 123,317,200 Yuan, RMB 235,947,500 Yuan, RMB 61,402,200 yuan, and RMB 13,795,700 yuan respectively for compensation of the basin. Under the target control on water quality in the horizontal compensation model, measured by environment capacity, Hebei Province should pay RMB 34,532,300 Yuan for the compensation of the basin. Henan Province should pay RMB 67,742,100 Yuan each to Hebei and Shandong Provinces as environment pollution compensation, while Shandong Province should pay RMB 85,372,800 Yuan to Hebei Province.

4.4.4 Rural Non-Point Source(NPS) Pollution

Research method: This study includes production sources such as farming, and livestock & poultry breeding, and rural domestic sources (rural living) in agricultural NPS pollution. Compared with point source pollution, agricultural NPS pollution has the following features:

Dispersion of pollutant emission: Contrary to the concentrated point source pollution, NPS pollution is dispersed. NPS pollution is characteristic of spatial and temporal heterogeneity because of the change of land use conditions in the basin, topography, hydrological features, climate and weather, etc.. The dispersed emission in turn leads to the difficulty in recognizing the polluting geographical boundaries and spatial locations.

Random occurrence: NPS pollution often happens when it rains. The randomness of precipitation and the uncertainty of other factors determine the randomness of NPS pollution. The emission time of NPS pollutants cannot be easily predicted.

Difficulty to define source areas: The occurrence, diversion and formation processes of agricultural NPS pollution are complicated. Varied pollutant sources in one region and their cross emission, coupled with the influence of geography, meteorology, and hydrology on pollutants' source, fate and effects make it hard to define the exact single pollutant source of one specific polluted water body, or of the quantity and types of pollutants that come into the receiving water body.

The cumulative nature of the polluting process: When the chemical fertilizers and pesticides are applied to the farmland, the unabsorbed part or those that aren't degraded will exert an influence only when it rains. However, if there is rainfall or irrigation at the time when the chemicals have just been applied, then a serious NPS pollution may follow. Even when there is no precipitation and other ground runoff when the fertilizers are applied, they accumulate in the soil and can leach into the sub-soil and ground water with rainfall or irrigation water over longer periods of time.

Based on the above facts, it is to conclude that to exactly evaluate the agricultural NPS pollution load is extremely difficult, especially when China lacks abundant long-term observation data. Thus, it's necessary to develop a new method for evaluating the agricultural NPS pollution load along the basin with insufficient information.

There are three necessary conditions for the occurrence of agricultural NPS pollution: (i) the potential number of polluting factors; (ii) the ratio of pollutants transported to watercourses to the total pollutants; (iii) runoff. When more polluting factors exist, or the ratio is greater, or the runoff is large enough, the agricultural NPS pollution load is larger. Their relationships can be expressed in the following formula:

$$Q_{PL} = Q_{PI} \times C_i \times C_r$$

Here Q_{PL} refers to pollution load; Q_{PI} refers to the number of polluting factors; C_i refers to the ratio of the pollutants that flow into river, expressed as a coefficient; C_r refers to runoff ($\leq$ the runoff when all pollutants flow into river).

The potential number of polluting factors is an important factor of agricultural NPS pollution. The causes of NPS pollution are varied in different agricultural production fields. For example, the major source for pollution in farming comes mainly from the overuse of nitrogen and phosphorus fertilizers on farmland, while the major contributor in livestock feeding is the inappropriate management of excrement of livestock and poultry.Some of the nitrogen and phosphorus nutrients applied in the farmland are absorbed by crops, and a small amount of them evaporate into the atmosphere, while others become lost in the soil and become potential pollution sources. According to the diversion and transformation mechanism of nitrogen and phosphorus nutrients, the pollution load in farming is calculated as follows:

The potential pollution load in farming = (before fertilization) ± the amount of nitrogen and phosphorus in the soil + the application amount of nitrogen and phosphorus in that year − amount absorbed by crops − amount lost in vapor (volatilization).

During the livestock feeding period, unprocessed livestock wastes are often discharged into water bodies and cause contamination. Since, under certain technical conditions, the excretion amount urinated by the same kind of livestock is almost similar during its growth till it's old enough to be killed, the potential pollution load in livestock can be evaluated with different livestock feeding amount, feeding cycle, and daily or yearly excretion coefficient, the amount of excretion with innocent treatment, and the average amount of the pollutants (TN, TP, NH_3-N, COD) in different kinds of livestock, as is shown in the following expression:

Potential pollution load in livestock feeding = (the amount of livestock fed × daily excretion coefficient × feeding days − amount been treated properly) × average pollutant amount in the livestock excretion

Domestic pollution in the rural daily life comes mainly from the daily consumption and physical waste of the residents living in the rural areas. In certain life conditions, the daily domestic discharge per head and pollutants contained can be investigated to be determined and used to evaluate the potential domestic pollution load. The formula is shown below:

Potential domestic pollution load = permanent resident population × the daily coefficient of the pollutants produced per person × 365 days × (1-percentage processed)

In both formulas, the calculation is for "potential" NPS pollution load insofar as these formulas do not include the transport coefficient the connects load at source, with load that is actually delivered to nearby watercourse. In this analysis the actual pollution load to watercourses is not calculated because the transport coefficient is not known. Further information on this can be found in Yang et al., 2012.

During the evaluation process the county is used as the basic unit as there is statistical data from relative departments. Field surveys were carried out to understand the agricultural production and typical conditions of Hai River Basin, such as population, average daily expense, crop structure (including crop varieties, crop yield, and planting area), nitrogen and phosphorus fertilizer amount, and livestock structure (such as animal varieties, and number of animals fed), and to become familiar with the detailed condition of agricultural NPS pollutionsources.

This method can be applied with statistical data collected by various departments (agriculture, water resources, population department, etc.) and with the revised technical parameters collected through field research in specific basin or regions to evaluate the potential agricultural NPS pollution load. As this method combines field survey with relative statistical data, it is referred here as the "IS" (Investigation-Statistics) Agricultural NPS Pollution Capacity Estimation. With ISAgricultural NPS Pollution Capacity Estimation, the potential agricultural NPS pollution load in Hai River Basin can be evaluated. In the evaluation, the livestock excretion coefficient and excretion pollutant coefficient are properly adjusted according to the agricultural production and agricultural pollution features in the basin. Because of the lack of current data on nitrogen and phosphorus content in the soil of farmlands in the basin, it is assumed that excluding losses to surface and groundwater and volatile loss of urea-N, the rest of the nutrients are absorbed by crops and become part of the soil nutrients of the farmland.

Table 4.7 Evaluation of Potential Agricultural NPS Pollution Load in Hai River Basin in 2007

units = tons

Pollutants	COD	TN	TP	NH_3-N
Farming (crops)	—	4,537,595	374,647	268,897
Livestock & Poultry	1,222,579	66,851	17,037	32,948
Daily Life	3,401	279	32	133
Atmospheric Deposition	—	3,670	—	—
Total	**1,225,980**	**4,608,395**	**391,716**	**301,978**

Major results: Using national census data of polluting sources from 2007, the agricultural NPS coefficient and the total potential polluting load are defined. The results are shown in Table 4.7.

The evaluation results show that in the potential agricultural NPS pollution load in Hai River Basin, COD mainly comes from livestock (99.72%) with a minor amount from daily life. For TN, farming accounts for 98.46%, livestock & poultry for 1.45%, daily life for 0.01%, and atmospheric sedimentation for 0.08%. For TP pollution, farming is the major source, accounting for 95.64%. In NH_3-N pollution, farming accounts for 89.05% and livestock & poultry for 10.91%.

Hai River Basin consists of many sub-basins, each with different agricultural

NPS pollution potential. Thus, partitioning by sub-basin is necessary to establish their polluting potential. This is best expressed as potential pollution per unit of area so that basin area does not distort the comparison by sub-basin. Table 4.8 shows sub-basins' potential agricultural NPS load per catchment area (including crops and livestock & poultry).Table 4.8 indicates that Cetian Reservoir in Yongding River has the least potential NPS pollution, followed by the downstream plain in North Sanhe River and North Sihe River. The most seriously polluted area, Xiajia River in Tuhaima is the key target area for controlling the agricultural NPS pollution.

Main countermeasures & remedies: As the important food supplier for many large and medium-sized cities (Beijing, Tianjin, etc.) and as the major agricultural producing area, Hai River Basin's steady agricultural development is of great importance. A long-term plan is needed for the harmonious development of agricultural production, for rural economic and social development and agricultural NPS pollution control and prevention, is needed. The following summarizes the main recommendations.

- Agricultural emission reduction should be encouraged.
- The farming development mode and policies should take farming structure optimization as the lead, and realize the reasonable utilization of land, appropriate application of chemicals and fertilizers and water-saving irrigation with scientific farming methods.
- Straw should be recycled to improve its utilization rate, and develop a resources-products-recycle-reproduction cyclic development mode.
- Livestock & poultry development mode should start from the choice of the feeding sites, and gradually regulating NPS pollution in these sites with point source treatment methods.
- Biogas engineering projects should be carried out for the recycling of livestock excretion as the feedstock, and efficient production and application technologies of organic fertilizer should be promoted.
- Finally, a cyclic agricultural mode combining breeding and growing should be developed so that all waste is used efficiently for soil improvement and for energy production (biogas).
- Human waste from rural daily living requires new approaches to waste management, including consideration given to centralized and scattered villages

according to the features of those villages. In urbanized and centralized villages, measure are taken through civil engineering, while in scattered villages, simple, low-cost and effective treatment methods are preferred to reduce NPS pollution.

Table 4.8 Unit Potential Pollution Load of the Farming Areas in Sub-basins in 2007

unit: tons/km^2

Sub-basin areas	Catchment Area/ km^2	COD	TN	TP	NH$_3$-N
The Whole Basin	569,338	2.14	8.10	0.70	0.54
Mountainous Region of North Sanhe River and Downstream Plain in North Sihe River	83,443	0.84	2.94	0.29	0.18
Cetian Reservoir in Yongding River	108,690	0.73	2.02	0.22	0.12
Daqinghe Basin	166,917	0.87	3.96	0.35	0.24
Heilong Harbor and Yundong Plain	31,487	2.04	17.94	1.93	1.05
Plain and Mountainous Areas in Zhangwei River	58,284	4.18	13.53	1.00	0.90
Plain and Mountainous Areas in Ziya River	17,654	6.08	30.26	2.38	2.00
Luanhe Plain and Mountainous Areas & Rivers along Eastern Hebei	77,436	2.08	4.29	0.42	0.37
Xiajia River in Tuhai	25,427	13.79	49.65	3.72	3.55

4.5 DEMONSTRATION PROJECTS

To better implement IWEMPs in the 16 pilot counties (cities, or districts) demonstration projects were developed to highlight methods to solve typical water resources and environment problems. These fall into three main categories: water resources management, pollution control, and ecological management.

4.5.1 Demonstration Projects of Water Resources Management

4.5.1.1 "Water Conservation" Demonstration Project in Beijing

"Technical Study on Water Conservation with Remote Sensing ET Data in Beijing"is a core topic of water resources demonstration programs. Its overall objective is to improve agricultural water management capability in irrigation areas, realize the transformation from "water supply management" to "water demand management", and achieve source water conservation and sustainable usage of water resources. This topic study is based on ET remote sensing survey system in Beijing, and by combined adoption of theoretical study, survey data analyses and model calculation. It applies remote sensing ET data in agricultural water management with remote sensing ET temporal and spatial data, land usage, and crop distribution generated by the system, and the field survey documents of crop yields and precipitation, combined with observed data of regional water consumptions and ground ET survey data of different regions.

The topic studies mainly the following three aspects: (i) remote sensing of ET based regional water balance analyses and prediction; (ii) regional ET quota and irrigation water quota allocation; and (iii) irrigation management and evaluation of groundwater irrigation areas. The technical route of the topic is shown in Figure 4.4.

The topic has achieved the following findings:

  The water consumption pattern of main crops and spatial distribution of moisture generating rates are studied based on the remote sensing survey of regional water consumption and water generation distribution, and a regional crop moisture generating function is constructed. An RS based model is used to compute crop water consumption quota and irrigation water quota is proposed for the first time, which provides an effective way for water consumption quota management in areas with a deficit of water resources.

  A water balance model is used to analyze the supply, demand, expense, and draining conditions in the studied areas both in current and future situations; it provides evaluations and predictions under different situations such as when improving irrigation water utilization efficiency, when practicing deficit irrigation, when improving the farming structure, and when utilizing water resources outside the region. It makes up the defects of the traditional supply and demand balance method, and is a valuable reference for farming structure optimizing and reasonable allocation and efficient utilization in the region.

  Based on the long-term observation data of groundwater data and RS ET data, a quantitative relationship between water consumption in the region and groundwater change range is established. Different water-saving measures' effects upon reducing regional water consumption and cultivating groundwater are also

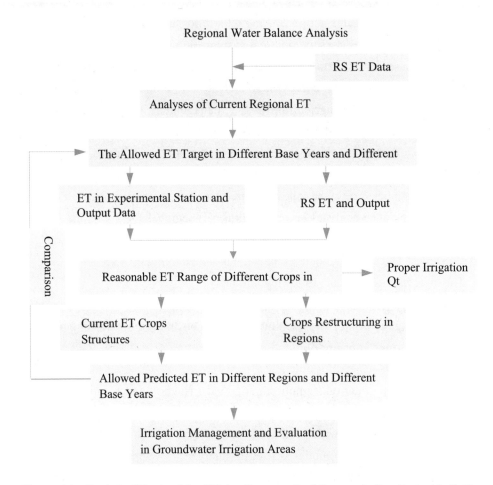

Figure 4.4 Technical Route of the "Water Conservation" Demonstration Project in Beijing

studied, which a offers theoretical basis for agricultural water-saving decision-making. The application of RS ET in agricultural water management is enabled.

In this project, Daxing District and Miyun County are chosen as the demonstration area with both plains and mountains. By discovering and solving the problems existing in agricultural water management in those two areas, RS ET based water utilization management tools and methods are put forward. Those tools have already been installed in their water management system and applied in practical agricultural water management in the demonstration areas. With the major crops ET quota and irrigation water quota obtained through the tools, quota management of water consumptions of main crops is practiced, which in turn could save the exploitation of almost 20 million m³ groundwater, and realize regional source water conservation. This study also puts forward regional water balance analysis tools and irrigation water management evaluation methods based on regional water consumption control. With those, the real regional water-saving requirements can be clearly identified, and corresponding advanced water-saving measures can be adopted which can not only relieve the local water shortage, but also provide guidance of how to effectively utilize water resources for similar areas with severe source water shortage.

4.5.1.2 "Water Right Allocation and Groundwater Management" Project in Guantao County, Hebei Province

Located in southern Hebei Province, and upstream of Heilong Habour of Haihe River Basin, Guantao County is in the warm temperate semi humid climate zone, with an average precipitation of 548.7 mm for many years. It has no imported (water transfer) water. With a land of 456.3 km² and an arable land of 467,000 mu, Guantao County has 62.6 million m³ water resources in total, namely, 134 m³ per mu, and 196 m³ per person, which is far below the average of the province and that of the whole country. It is a place with grave water resources shortage. Thus, this project aims to explore specifically how to realize demand and supply balance of water resources, and the sustainable development of underground water environment and zero over-exploitation of ground water in such regions.

Relationship between ET and water rights: Some elements of ET and related management modes had been introduced to the groundwater-well irrigation area in Phase I of the earlier Water Conservation and Irrigation Project carried out under a loan from the World Bank. In this GEF Hai River Project, it is the first time to introduce water rights to the well irrigation area and advanced ET management methods. ET is an abstract concept, which is hard for farmers to understand. However, as agricultural water consumption accounts for more than 80% of the total water consumption in the county and a significant amount comes from groundwater, it is the focus for ground water management.

Rainfall that infiltrates as ground water is regarded as controllable amount of water and is allocated to towns, villages, and households (plots) as "water right". Crop $ET_{single} = P_{effeciency} + W_{irrigation}$ is also controllable water amount- water right amount; ET_{arable} = the weighted average of each $ET_{single} \times$ farming area; $ET_{non\text{-}arable} = 0.6 \times ET_{arable}$; $ET_{total} = ET_{arable} \times \eta + ET_{non\text{-}arable} \times (1-\eta)$. From the above relationships among those variables, it can be seen that the control of $W_{irrigation}$ is essential to the control of ET_{total}.

- ET_{single} — each single crop's ET value;
- ET_{arable} — ET value of arable zones;
- $ET_{non\text{-}arable}$ — ET value of non-arable zones;
- ET_{total} — the total ET value of arable zones and non-arable zones;
- H — the efficient utilization coefficient of the arable zones;
- $W_{irrigation}$ — ground water exploitation amount.

The determination of ET_{total} value: Guantao County has a long-term average precipitation of 548.7 mm; the annual average for 1985-2005 used in this GEF Hai River Project is 530.2 mm. The allocated ET quota is 556 mm. After comprehensive analyses, the average precipitation is taken as the target ET value. Consequently, ET ~ annual water consumption ~ total annual rainfall. Years with abundant rainfall can complement dry years, mainly through groundwater storage. In principle, water demand is then satisfied by rainfall, and the water resources' inflow and outflow can be balanced.

Water right allocation objective: The objective of the water rights allocation is to replace the traditional management with an operational ET management model, which is based on ET as its water right allocation standard, and implement engineering water-saving technologies, and agricultural water-saving and water-saving management measures. It also aims to realize zero ground water excessive exploitation, and make the annual total ET value equal to the local average annual precipitation. The objective is also to guarantee the continuous development of agriculture and the national economy as well as the sustainability of rural society.

Water right allocation methods: The allocation of water right is government responsibility. With a re-evaluation of the whole county's water resources, the agricultural water is allocated based upon accurate calculation of average ET in arable zones, the ET_{total} of both the arable and the non-arable zones, and the allowed average exploitation amount of the ground water. The specific steps are as follows:

- According to the hydrological types of the sub-regions' ground water in the county, the ET_{arable}, ET_{total}, and the corresponding allowable exploitation amount are calculated.

- Those sub-regions are then furthered divided into smaller sub-partitions, and their ET_{arable}, ET_{total}, and the correspondent allowed exploitation amounts are calculated respectively.
- Villages are classified as one sub-partition, and their ET_{arable}, ET_{total}, and the correspondent allowable exploitation amounts are calculated.
- Each village allocates their ET_{arable} and allowed exploitation amounts to each household (plot) according to the contracted areas of individual farmers (the contractors).
- Non-arable ET is controlled by the village, and is mainly used for daily water consumption of people and livestock and other public welfare undertakings.
- In areas where groundwater is excessively exploited, ground water should be exploited and utilized in a rational way. When ground water shortage occurs, each farmer should share the shortage part.

Water right allocation outcome: With nearly four years' analyses and calculations, the technical staff in Guantao County PMO and county water resources office have allocated the groundwater rights to each village and town, and further from towns and villages to each household. The details are shown in Table 4.9.

Table 4.9 Water Rights and Resources Rllocation for Guantao County

Name of Towns/Villages	Area of Land/ km²	Arable Lands/ hm²	ET_{arable}/ mm	Non-arable ET/mm	ET_{total}/ mm	Allowed Exploitation Amount/10,000 m³		
						Sub-Total	Arable Lands	Non-arable Lands
Guantao Town	50.72	2,025.24	698.8	415.7	526.55	667.7	267.4	400.3
Weisengzhai Town	55.98	3,169.19	665.7	399.4	551.2	771.4	437.3	334.1
Luqiao Village	72.26	4,120.91	665.9	399.5	551.4	996.1	568.8	427.3
Shoushansi Village	60.39	3,168.88	692.9	415.7	559.86	845.2	444.0	401.2
Wangqiao Village	56.01	3,206.86	656	393.6	543.2	760.6	436.3	324.3
Nanxucun Village	42.69	2,486.13	637.3	382.4	530.2	565.9	329.6	236.3
Fangzhai Town	43.78	2,611.83	659.8	395.9	554.2	606.6	361.9	244.7
Chaibao Town	74.47	4,468.01	669.2	401.5	562.14	1,046.6	630.7	415.9
Total	456.30	25,265.53	669.0	401.4	548.7	6,260	3,476.0	2,784.1

Water rights system: Without water rights system, water resources are difficult to allocate. Therefore, water rights system is a reliable guarantee for water resources allocation. According to Guantao County's actual situation, the water rights implementation principles are as follows:

- On behalf of the State Council and local government, the Guantao County People's Government manages the ground water resources within its boundary and fully coordinates the resource allocation.
- Land owners have the usage right (or relative ownership) of the ground water resources covered by the lands he owns, including its priority right.
- Water resource usage right will be changed or transferred when the land usage objective changes.
- Any exploitation and waste of ground water resources that disregards national legal constraints or the interests of other users should pay for any consequence loss.

Management of Water Rights System:

Management System: The county Water Resources Management Committee is responsible for implementing water resources laws and management. The committee is composed of water resources management office and two law enforcement units, responsible for executing relative laws and regulations. Apart from abiding by the national and the higher local people's government's laws and regulations, Guantao also drafts its own documents such as Guantao Ground Water Management Plan.

Management Mechanism: Under the comprehensive management of the county water resources management committee, relevant organizations and departments concerned should cooperate with each other, constituting the water right management system and operating mechanism.

Regulation Construction: Water Resources Allocation and Water Rights System Construction Plan, Water Price Authorization and Water Fee Collecting System, IWEMPs Plan, Guantao GEF Demonstration Project-Further Ground Water Management Plan, Guantao Brackish Water and Salt Water Development and Utilization Technology and Management Guidance, and the Guantao County Ground Water Management Plan are developed and then transmitted for approval by the county People's Congress.

Ground Water Management Outcomes and Achievements: Since 2000 when Guantao was chosen as the demonstration county for the WCP ground water management loan of the World Bank, groundwater management has made great progress. Now, as the groundwater management demonstration project in GEF Hai River Project, Guantao County has achieved important mechanisms for its sustainable development of its ground water resources. Guantao's groundwater management system now not only witnesses sustainable development, but also is more comprehensive, systematic, and scientific.

The groundwater management has achieved obvious effects, and its over-exploitation rate has slowed down rapidly. From 2000 to 2009, the ground water decreases 0.21 m in average, a decrease of 0.52 m from the average 0.73 m before 2000. Since GEF Hai River Project, the annual decrease rate achieved 0.18 m during the Five-Year period from 2005 to 2009 with 2004 as the base year. Excessive groundwater use for agricultural irrigation has decreased by 50.1% from the base line of 25.23 million m^3 to 11.401 million m^3. All those achievements (details in Table 4.10) show the effectiveness of the GEF Hai River Project, which improves further Guantao County's water resources and environment, and makes zero ground water exploitation and the sustainable utilization of water possible.

Table 4.10   The Average Depth of Ground Water in Shallow Aquifers   unit: m

Year	1999	2000	2001	2002	2003	2004	2005	2006	2007	2008	2009	Annual Average Decrease Rate
Guantao County	19.69	19.7	19.87	21.06	20.88	20.83	20.66	21.8	21.36	21.1	20.95	0.126
Demonstration Areas	—	—	—	—	—	—	20.3	19.9	21.1	20.9	21	20.7	0.08

Summary: (This is the first time in the North China Plain, and in groundwater well irrigation areas in the Hai River Basin, an ET-based system for managing groundwater and for control of its exploitation, and the use of water rights, has been applied.)

4.5.2   Water Pollution Control Demonstration Projects

4.5.2.1   "Water Pollution Control" Demonstration Project in Xinxiang County, Henan Province

In recent years, with the rapid economic and social development of Xinxiang County of Henan Province, population growth and urbanization is rapidly increasing together with increasing water consumption; this has resulted in excessive groundwater exploitation. Because of the irrational industrial structure, the county's pollution producing enterprises such as paper-making, are located mainly along Mengjiangnv River. The results in highly centralized pollution discharges. As a result, the river has lost its ecological functions, and the groundwater and the downstream have been severely polluted. This produces large loads to the Wei River River. Currently, the water supply situation in the county is grim and the water environment is rapidly deteriorating. In order to solve the water pollution problem in the Zhangweinan Canal (the east and west of Mengjiangnv River) of Xinxiang County, there is need to effectively control water pollution and protect the quality of water resources. For this reason a demonstration project was carried out in Xinxiang County to recommend a suite of interventions to the county government.

Industrial Pollution Control:

The following interventions are required for industries:

Raising Industrial Pollutant Emission Standards and Optimizing Industrial Structure: Based the existing emission discharge standards, and combined with the economic and environmental benefits of industrial development, strict emission limits standards were developed to control pollutant emission. Areas having local standards or industrial standards should apply them accordingly, and other areas should abide by the integrated pollutant emission standards of Xinxiang County. Taking into account Xinxiang's industrial development and resulting pollutant emissions, this project adjusted mainly those for industries with high water consumption and pollution such as paper-making, pharmaceuticals and medicine, and chemical companies. It also proposes to strengthen supervision management, practice more severe punishment on enterprises that discharge pollutants secretly or carelessly, or exceed the pollutant emission limits, and recommend suitable advanced technologies for water conservation and pollutant reduction.

The Development of Recycling Economy and Promotion of Clean Production: The project introduced the recycling economy and green GDP concepts, advocated sustainable production and consumption behavior, and promoted the recycling of resources. It aimed to realize energy conservation, consumption reduction, pollutant decrease, and efficiency improvement through such measures as transformation of the production processes, and technical enhancement of production methods, increases in the recycling rate of industrial water, and decreased unit sewage and pollutant emission per unit of industrial output.

Domestic Pollution Control:

The following interventions are required for domestic pollution control:

Sewage Treatment Plants Construction and Central Sewage Treatment Rate: To satisfy Xinxiang County's industrial sewage and domestic sewage treatment, to reduce industrial and domestic pollution, and to decrease pollution load of water bodies, especially those in the east and west of Mengjiangnv River, new sewage treatment plants are recommended in areas with heavy industrial and domestic pollution load and for those without treatment facilities. Also, supervision of treatment plants and operational management and maintenance of those treatment systems should be strengthened. All plants should install on-line monitoring devices to ensure that emissions meet the legal standards.

Speeding Up the Construction of Sewage Pipe Network of all Treatment Plants: Further improvement is needed for the water collecting pipe network for both industrial and domestic treatment plants. The network should be expanded and the processing rate needs improvement. All domestic sewage within the plants' collection area should be routed into these plants for treatment. All industrial sewage that can be treated in the plants should first be pre-processed to reach the emission standards required of individual enterprises, then sent to the treatment plants for further processing.

Other Measures Recommended

A rural sewage collection and treatment system should be established. For those villages that cannot be included in urban sewage treatment plants, waste treatment can be adapted to local conditions such as wetland treatment, forestry treatment, land treatment and other ecological treatment methods.

Newly built, renovated and expansion of buildings and residential communities should designed and constructed with separated waste systems so that rain runoff and sewage are separated. Before the regional central sewage treatment system is built, construction projects with large domestic sewage discharge should be equipped with domestic sewage treatment facilities. Those projects include hotels and entertainment facilities with daily discharges above 60 m^3, residential communities with daily discharge above 100 m^3, and multiple-functional buildings and apartments with daily discharge above 100 m^3.

Livestock and Poultry Pollution Control: According to Xinxiang County's livestock and poultry discharge reduction requirements, by the end of 2010, 80% of large-scale livestock and poultry breeding enterprises should install sewage treatment facilities and forbidden to discharge sewage directly into water bodies. Livestock excretion recycling rate should be no less than 90%. By 2015, 90% of the large-scaled livestock breeding farms and farming districts should be equipped with complete solid waste and sewage treatment facilities, and guarantee their satisfactory operation. The livestock excretion recycling rate should be no less than 95%. By 2020, all farms and farming districts should be equipped with solid waste and sewage treatment facilities, and the livestock excretion recycling rate should reach 100%.

River Treatment Scheme: The treatment scheme focuses on the east and west of Mengjiangnv River, People's Victory Channel, and Communism Channels. In the east and west of Mengjiangnv River, the pipe network should be improved and sewage collected into central plants for treatment. Comprehensive water environment treatment project will be carried out in the county and sediment dredging and water ecology restoration implemented to eliminate the occurrence of ClassV water bodies with attendant black water and objectionable odour.

4.5.2.2 "Water Pollution Control" Demonstration Project in Lucheng City, Shanxi Province

In IWEMP "Water Pollution Control" demonstration project in Lucheng City, two sub-projects were chosen as demonstration projects: (i) Artificial Wetland Water Purification Project in Nanyuan of Zhhuzhang River; (ii) the Ecological Compensation Mechanism of Water Resources Areas in Xin'an Spring, Shanxi Province.

Artificial Wetland Water Purification Project in Nanyuan of Zhuzhang River: Artificial wetlands have a large buffering capacity and can have excellent purification effects if well maintained. When combined with together water pollution control measures already undertaken in Lucheng City, artificial wetlands can achieve a reduction in ammonia nitrogen and to achieve water quality that can attain Class III in Zhuzhang River. An artificial wetlands water quality purification project in Nanyuan, Zhuzhang River was established in the upstream area of Zhuzhang River. The Nanyuan beach in Zhuzhang River and shallow water areas from North Song Village, Han Village to Dongbaitu Village in Dianshang Town were chosen as the construction areas. The construction area is about 1,400 m long, about 256 m wide, with a construction area approximately 540 mu (Figure 4.5).

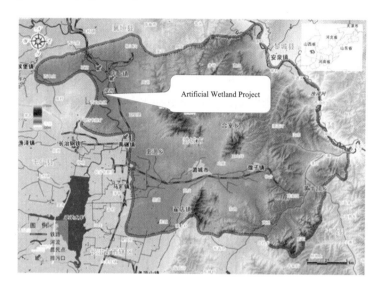

Figure 4.5　Zhangwei River Artificial Wetlands Water Purification Project Location

The project is based on: the design principles of stable technology, feasible economics and simple management; the topography of the construction site; taking into consideration the coordination between water purification and ecological protection; and the equal importance of environment benefits and economic benefits.

The artificial wetland water purification project plan was determined as follows: firstly, on the Nanyuan River and Zhuzhang river course, 400 m north of Song Village Bridge in Dianshang Town, a rubber dam was built to raise the river level. Upstream of the rubber dam, an ecological retention pond was formed to pre-process the polluted water in the river. Then, under the gravity, the water flows into the artificial wetland system where it undergoes biological purification through multi-level flows in the system and the combined effects of plants, microorganisms, and soil. With such purification, the water meets the Class IV emission standard and is allowed to be discharged.

The processed water can be used for landscape irrigation and farmland irrigation, or for covering the ecological water in the river. Thus, the recycling rate of water resources is enhanced, and both economic and social benefits and environmental benefits are achieved. In the long-run planning, artificial wetland Phase II is planned for Pai Village, the downstream of Nanyuan of Zhuzhang River to reprocess water and further improve the water quality.

Ecological Compensation Mechanism of Water Resources Areas in Xin'an Spring: Since Xin'an Spring's source water has been severely restricted, the water ecology has been destroyed and the water body polluted. Therefore, an ecological compensation mechanism sub-project was established to research such issues as scope of compensation, compensation criteria, compensation mode, and compensation implementation. This mechanism can help restrict and constrain unreasonable development and utilization of water and land resources, encourage ecological environment construction and protection activities, which are of great significance to restore the ecological environment of Xin'an Spring, and realize the sustainable development of the region's economy.

The compensation object includes all the residents who have made a contribution or sacrifice for the protection and restoration of the water ecology; the compensation subject are the units and individuals in the downstream areas who benefit from this project, including all individuals, enterprises, and departments that discharge pollutants in daily life or during production, and influence the water quantity and quality of the basin.

The compensation criteria are the total cost of ecological construction in water source sites during all kinds of production and daily life. It includes water conservation forest establishment and nursing costs, the project's construction costs, operation costs, economic loss caused by occupying farming lands and housing land, opportunity costs, etc..

The ecological compensation cost for Xin'an Spring Water Sources should be between the total cost and the ecological service value. Based on the data collection and field survey of the protection areas in Xin'an Spring, ecological compensation should be calculated according to the ecological compensation criteria in water source sites. The compensation willingness of people who are related with the interests involved in the project in the protection region, and the outcomes of protection should be investigated and analyzed. With all those above, the allocation of ecological compensation is finally determined. Results show that the ecological compensation for water source sites of

Xin'an Spring in Lucheng City is RMB 22.9295 million Yuan.

Finally, the ecological compensation management mechanism and suggestions about the Xin'an Spring were put forward, including water source sites protection ecological compensation charge policies, fiscal policies, investment policies, tax and support policies, and legislative recommendations or suggestions of Xin'an Spring ecological compensation.

Ecological Restoration Demonstration Project in Dezhou City, Shandong Province: Located in the middle reach of Zhangweinan Sub-basin in Haihe, the downtown of Dezhou is the control hub for upstream flood storage, and is the intersection of South–North Water Transfer Project. It not only is the area for sewage transit of three upstream provinces but also produces sewage from its own area. As a representative polluted reach of Zhangweinan Sub-basin, it is a key area to demonstrate restoration ecology that could apply to the whole basin.

In Dezhou, the sewage collection system is defective with only about 60% of the sewage in the city flowing into the pipe network. Large amounts of untreated sewage flows directly into the river and results in serious pollution. The discharged sewage also poses a threat to surface waters because of the defective sewage treatment and lack of advanced treatment. Groundwater is polluted directly or indirectly by these conditions. Extreme pollution results in water shortage and causes deterioration of the water environment.

The Zhangweinan Sewage Emission Project in Dezhou downtown started at the end of 1994, and was adjusted in 2004, after whichits treatment scale is 100,000 m^3/day. The treated sewage was then discharged through Xuanhui River to Cha River, and finally flowed into Zhangweixin River. Since the treatment plant was built long ago, its design did not include any specific technical requirements for nitrogen and phosphorus reduction. Thus, the incapability of denitrification and phosphorus reduction are serious deficiencies in the pollution control objectives for Dezhou. To satisfy the water demand for the city's development, to alleviate the current water supply shortage, and to reduce ground water exploitation and protect the limited ground water resources, and to guarantee the implementation and practice of sustainable development, the Dezhou Government decided to carry out a program of advanced sewage treatment and wastewater recycling. This project aims to renovate the existing plant with a treatment scale of 100,000 m^3/day, add denitrification and phosphorus reduction functions, and add a water recycling program with a scale of 40,000 m³/day, with the recycled water used for urban greening.

The south component of the comprehensive pollution treatment project of Dezhou Downtown is comprised of three parts: sewage collection pipe network construction in South Canal (it mainly refers to improvement of the existing pipe network, establishment of Xiaozhuang Pumping Station, and strengthening of sewage collection capability), design adjustment of existing projects, and reclaimed water recycling project. Specifically, design adjustment of existing projects means restructuring with the existing ponds, including the restructuring of anaerobic ponds, the denitrification and dephosphorization functions of facultative ponds, and improvement of existing pipe network. Reclaimed water recycling project refers to advanced treatment of reclaimed water flowing from the plant and the construction of Xiaozhuang Pumping Station.

Sewage Collection Pipe Network in South Canal: Collect sewage from the economic development zone in the canal and put it into the original North Plant Pumping Station built in the sewage emission project in Zhangweinan Sub-basin through Xiaozhuang Sewage Pumping Station and water pipes. After another central treatment in oxidation ponds, it meets the standard and is recycled or used for irrigation. The rest is discharged through Nangan Channel from Fen River to the downstream.

Project Restructuring Design: Based on the current sewage treatment plants' inflow water quality, operational conditions, and treatment cell conditions, establishment and renovation of the original two anaerobic ponds with A/O technology. At the head end of the ponds, a slow sand filtration system is built to further process sewage by removing SS, so that the final treated water can meet the first class-B level clarified in Sewage Emission Standards of Treatment Plants in Towns (GB 18918—2002).

Reclaimed-water Recycling Treatment Project: The sewage, after biochemical treatment, is further treated to remove the remaining contaminants with V-typed filtration ponds and disinfection process, in order to meet the users' requirements of reclaimed-water. To guarantee the inflow water can reach the first class-B level clarified in Sewage Emission Standards of Treatment Plants in Towns (GB 18918—2002), the slow filter is added in the head end of facultative ponds based on the renovation of anaerobic ponds through A/O technology, so that the outflowing water quality is stable.

Chapter 5
INTEGRATED WATER RESOURCES AND ENVIRONMENT MANAGEMENT PLANS (IWEMPs)

5.1 OVERVIEW OF INTEGRATED WATER RESOURCES AND ENVIRONMENT MANAGEMENT PLANNING

Integrated Water Resources and Environment Management Plans (IWEMPs) are the principal outcome of the GEF Hai River Project and are the basis for improving management of water resources and the water environment (especially pollution control) at the county (district) level and which, in turn, creates water savings that can be used both to reduce groundwater overdraft and to pass downstream to the Bohai Sea.The IWEMP planning process integrates technology, administration, economic, social and legal approaches, and includes the "top-down, bottom-up"interactions and cooperation. It includes public participation and wide stakeholder involvement to create a sense of ownership. The objective is to produce an IWEMP that implementable, sustainable and reproducible.

As an example, the IWEMPs of Beijing Municipality's demonstration counties (districts)closely interrelate remote sensing monitoring ET technology, strategic research outcomes that provide a scientific basis for integrated management,economic data and water pricing, modeled outcome scenarios. Using criteria of economic targets, ecological possibilities, and feasible ET reduction, the Plan uses supply-demand analysis and integration and analysis on possible measures of reasonably curbing water demand, effectively increasing water supply, and protecting the ecological environment. Each county/district IWEMP compares each feasible water resources allocation scheme and water pricing scheme and to arrive at a recommended scheme that will resolve the county (district) water resources and environment issues.

Each of the sixteen demonstration counties of his Project developed IWEMPs. For this monograph, we present these examples: (i) a major metropolitan area (Tianjin); (ii) a mountainous area (Pinggu district of Beijing Municipality), (iii) a small city (Xinxiang); and (iv) a predominantly rural county (Guantao).

5.2 TIANJIN MUNICIPALITY IWEMP

Tianjin Municipality is China's northern economic center and international port city and is part of a national development strategy. Tianjin Municipality is an area of severe water shortage, with major development and high utilization of water resources, and with large wastewater discharge that creates as series of ecological problems. Water resources are the limiting resource constraint on development of this area. Therefore, it is urgent for Tianjin Municipality to formulate one set of feasible integrated water resources and environment management policy and measures that can lead to economic and social sustainable development. The IWEMP approach is new to China and to Tianjin, therefore the Hai River Project used Tianjin as a major example of this methodology and of the benefits of close coordination of the environment and water sectors.

5.2.1 Tianjin Municipality Overview

5.2.1.1 Basin Situation of Tianjin Municipal Water Resources and Environment

Water Resources Overview: From 1956 to 2000, the total annual water resources of Tianjin Municipality is approximately 1.55 billion m^3, This is 160 m^3 per capita which is about 1/15 of national available water resources per capita. In 2004, the total water supply for Tianjin Municipality was 2.206 billion m^3 of which surface water was 1.489 billion m^3 and groundwater was 0.707 billion m^3. This large amount of water consumption causes discontinuous river flow, significant decrease of runoff into the sea and continuous decrease of groundwater level. Hence, unsustainable over-exploration of water resources is a major issue for this area.

Water Environment Overview: With large wastewater discharges, lack of sewage treatment capacity, and decrease of water body self-purification capacity, the municipality's water bodies are seriously polluted. Of the nineteen 1st Class Rivers in the municipality, the water quality of most are worse than Class V (Class V is the worst

Chapter 5 INTEGRATED WATER RESOURCES AND ENVIRONMENT MANAGEMENT PLANS (IWEMPs)

class of water on the Chinese water quality 5-point scale and is unfit for most beneficial purposes).

Water Ecology: From 2000, the groundwater annual exploration amount from whole municipality is over 0.7 billion m^3, over-exploration of deep groundwater in whole municipality is severe, with annual over-exploration amount over 0.4 billion m^3. There is no baseflow in most rivers due to drawdown of groundwater. Annual average number of dry days (zero discharge) of major rivers in TianjinMunicipality is 320 days. Wetlands area is about 650,000 mu (43,333 hectare), which is 80% less than that of the 1920s. Heavy pollutant discharge has also caused serious deterioration of Tianjin's water ecology.

Water resources and environment management: Typical of China, water-related issues management is allocated in various departments, with overlapping departmental power, excessive division of labor, and too many procedures; therefore, an integrated approach to water quality and quantity planning and management system has never been developed. The current water-related planning is defined by the relevant departments in accordance with their respective power scopes, with planning overlapping and conflict issues. Also there is a lack of sufficient public participation, excessive regulations and standards, and poor availability of information.

5.2.1.2 Significance of Tianjin Municipality IWEMP

The annual consumed ET from 1980 to 2004 inTianjin Municipality is 43.4 mm higher than the available ET, among which the annual consumed ET is 147.4 mm higher than the available ET after the year 2000. The over-consumed ET is obtained by over-extraction from surface and ground water, thereby reducing water transfer quantity to the sea. The solution lies in bringing regional integrated ET requirements down the level of available ET. Since the water issues in Tianjin Municipality are typically highlighted by ecological environment deterioration due to water resources over-exploration and large discharge of pollution, the water ecology and environment can be fundamentally improved only through combination of water resources and environment planning and management, and the combination of pollution reduction and flow volume increase.

5.2.2 Highlights of Tianjin Municipality's IWEM Plan

The IWEM Plan is based on legal, institutional and technical review of all aspects of water and environmental management within the municipality. The Plan also factors in the results of the special studies that were carried out (two of these are reported below) into critical issues such as water quality and ecological issues. The following describes the major directions included in the Tianjin IWEMP Plan.

5.2.2.1 Management Target

The overarching target is to establish an ET-based integrated water resources and environment management scheme that includes administrative, legal, economic and technology perspectives.

The comprehensive targets are:

- To improve water resources utilization efficiency.
- To achieve the real water-saving by controlling water consumption.
- To provide water resources support for economic-social development target.
- To establish a scientific basis for water resources development, utilization and protection structure.
- To rehabilitate and improve Tianjin Municipality water ecological environment.
- To make necessary contributions for improvement of Hai River Basin and Bohai Bay water ecological environment.
- To develop a preliminarily form of modern ET-based integrated water resources and environment management framework.
- To explore new management approach for water resources sustainable development in water resources scarcity and seriously-polluted areas.

Specific targets are based on the recommended plans from Tianjin Municipality Water Resources and Environment Integrated Planning (2010D, 2010G, 2020D and 2020H) and are classified into seven controlling indicators, as illustrated by Table 5.1. The role of the South-North Water Transfer Project are factored into these targets. Explanation of the targets is provided below and the specific targets are noted in Table 5.1 for 2010 and 2020.

Chapter 5 INTEGRATED WATER RESOURCES AND ENVIRONMENT MANAGEMENT PLANS (IWEMPs)

Table 5.1A 2010 Tianjin IWEMP Plan Management Specific Targets, 2010

	Seven Controlling Indicators	Plan 2010D (Half Transfer of Middle Route from South-North Water Transfer Project)				Plan 2010G (Non-transfer from Middle Route of South-North Water Transfer Project)			
		Annual Average	50%	75%	95%	Annual Average	50%	75%	95%
1	Surface Water Total Volume Control (100 million m^3)	26.13	29.92	27.73	24.19	25.07	29.05	26.31	21.33
2	Groundwater Total Volume Control (100 million m^3)	5.78	4.22	6.08	8.68	6.40	4.97	5.19	8.58
3	National Economic Water Consumption Total Volume Control (100 million m^3)	28.83	30.92	30.72	30.55	28.40	30.80	28.40	27.65
4	Ecological Water Consumption Total Volume Control (100 million m^3)	3.07	3.22	3.09	2.33	3.07	3.22	3.10	2.26
5	ET Total Volume Control (mm)	626				611			
6	Pollutants Discharge Total Volume Control (Ten Thousand Tons)	1.70(NH_3-N) / 13.20 (COD)							
7	Total Volume Control into Sea (100 million m^3)	13.76	16.43	6.04	5.00	13.65	14.89	4.86	4.77

Table 5.1B 2020 Tianjin Municipality IWEMP Plan Management Specific Targets, 2020

	Seven Controlling Indicators	Plan 2020D (Full Transfer from Middle Route, Half Transfer from East Route of South-North Water Transfer Project)				Plan 2020H (Full Transfer of Middle Route, Non-transfer of East Route from South-North Water Transfer Project)			
		Annual Average	50%	75%	95%	Annual Average	50%	75%	95%
1	Surface Water Total Volume Control (100 million m^3)	34.36	36.74	35.74	32.40	33.26	36.10	34.41	28.77
2	Groundwater Total Volume Control (100 million m^3)	3.89	4.38	3.55	5.08	4.08	4.36	3.46	5.22
3	National Economic Water Consumption Total Volume Control (100 million m^3)	32.83	35.78	34.49	33.60	32.67	35.45	33.17	30.58

		Plan 2020D (Full Transfer from Middle Route, Half Transfer from East Route of South-North Water Transfer Project)				Plan 2020H (Full Transfer of Middle Route, Non-transfer of East Route from South-North Water Transfer Project)			
4	Ecological Water Consumption Total Volume Control (100 million m^3)	5.42	5.34	4.80	3.88	4.66	5.01	4.71	3.40
5	ET Total Volume Control (mm)	655				635			
6	Pollutants Discharge Total Volume Control (Ten Thousand Tons)	1.03(NH$_3$-N) / 7.51 (COD)							
7	Total Volume Control into Sea (100 million m^3)	16.51	18.02	9.43	8.02	16.49	18.01	8.88	8.20

5.2.2.2 Integrated Water Resources and Environment Management Measures

In accordance with above management targets, Tianjin Municipality will implement the ET-based integrated water resources and environment management through the following specific measures:

- Reform management scheme and system, establish new management model.
- Comprehensively implement controllable ET management, achieve "real water-saving".
- Implement surface water total volume control, achieve regional surface water optimization allocation.
- Strengthen groundwater total volume control, achieve groundwater exploration-supply balance.
- Enhance pollution treatment, achieve water function zone compliance.
- Enhance aquatic ecological rehabilitation, improve aquatic ecological status.

5.2.2.3 Measures Taken

With advancement of Integrated Water Resources and Environment Management Drafting, the water resources and environment management work in Tianjin Municipality has achieved the following.

- Completed Water Affairs Management System Reform: Tianjin Municipality

Chapter 5 INTEGRATED WATER RESOURCES AND ENVIRONMENT MANAGEMENT PLANS (IWEMPs)

formed the Tianjin Water Affairs Bureau in May, 2010; this is sector of municipal government. The responsibilities of the former municipal Water Resources Bureau, water supply responsibility of municipal Construction Commission, urban discharge and river banks management responsibilities of the municipal Roads Bureau, are incorporated into the new municipal Water Affairs Bureau. The Tianjin Municipal Water Resources Bureau is abolished.

- Further Refinement of Water Resources and Environment Management System: In recent years, Tianjin Municipality has issued or modified Tianjin Municipality Implementation Measures of People's Republic of China Water Law, Tianjin Municipality Municipal Water Supply and Utilization Regulation, Tianjin Municipality Water Pollution Prevention and Treatment Management Measures and Tianjin Municipality Clean Production Promotion Regulation. These form the regulatory framework for water resources and environment management. The "Tianjin Municipality Local Integrated Sewage Discharge Standard" was also promulgate; it provides more stringent local standard than the national integrated sewage discharge standard. It lists five major pollutants discharge limit values, including COD and NH_3-N , and also six indicators, such as allowable maximum discharge water volume for industry.

- Stricter Groundwater Resource Management: The municipality government has defined a groundwater exploration exclusion zone in 2007. The Municipal Water Affairs Bureau, in accordance with management targets of exclusion and restricted areas, redefined water-taking permitted volume for groundwater users, reduced exploration amount by 100 million m^3 within several years in the whole city, increased the groundwater resources fee standard, implemented stricter water-taking permission management, and strengthening compliance assessment.

- Launch Water Environment Treatment Three Year Action Plan: To effectively improve the water environment, from 2008 Tianjin Municipality launched a water environment treatment three years action plan to treat 39 seriously-polluted rivers, including block of sewage outlets alongside river banks, establishment of relevant discharge networks, and the renewal, reconstruction and expansion of sewage treatment plants. Currently, treatment of the 10 rivers within the central urban area have been mainly completed and the treatment of 29 rivers surrounding central

urban area have been partially completed.
- Advancement of Agricultural Water Utilization Management: Water Users Associations have been established in the whole municipality to encourage farmer to manage water themselves. Other measures include provision of measuring facilities in a large irrigation zone water-conservation reconstruction project, exploration of charging systems that reflect water volume use and to improve the water-conservation enthusiasm of framers,developing urban style water conservation agriculture and optimizing and adjusting agricultural cropping structure. To the end of 2008, the water-conservation irrigation area in the municipality reached 228,666 hectare (3.43 million mu), accounting for 65.8% of the effective irrigation area; the irrigation water utilization coefficient rate reached 0.63; the facility-assisted agricultural area has been developed for 36,667 hectare (550,000 mu).

5.2.2.4 Beneficial Experience

The IWEMP process has led to a number of valuable lessons based on beneficial experiences in this Project.

- Scientific Concepts and Advanced Technology are Required to Support Sustainable Water Resources Utilization: ET-base integrated water resources and environment management, with the support of advanced technologies such as remote monitoring ET, has established an ET-based integrated water resources and environment planning model and provides feasible integrated water resources and a solid environment management planning scheme.
- "Top-down, Bottom-up" Concept is Important Guarantee for Smooth Implementation of IWEMP: At the beginning of the Tianjin Municipality IWEMP project, the Municipal Bureau of Finance reported to the municipal government about the project, which then received support and recognition from municipal leaders. During project implementation, leaders from the water and environment sectors paid much attention to progress and provide guidance that enabled new and significant measures such as joint water resources and environment protection water function zoning, sharing of data, joint planning of outcomes from research topics, and smooth implementation of water conservation and pollution treatment projects. Governments, others stakeholders and research institutions cooperated closely, to establish a complete system from counties (districts) to cities, from

Chapter 5 INTEGRATED WATER RESOURCES AND ENVIRONMENT MANAGEMENT PLANS (IWEMPs)

special planning to integrated planning, which greatly ensures the smooth drafting and implementation of planning. "Bottom-up" principles included responsibility for analysis and planning at local levels.

- A Prerequisite Conditions for IWEMP Implementation is to Further Refine Water Resources and Environment Protection Cooperation Mechanism: This IWEMP is the first step for municipal integrated water resources and environment management.The advancement of integrated water resources and environment management and effective implementation of Tianjin Municipality IWEMP will only be fully realized with continuous refinement and advancement of water resources and environment protection cooperation mechanisms.

5.3 TIANJIN MUNICIPALITY IWEMP SPECIAL STUDIES

The Tianjin Municipality IWEMP was developed though six, inter-related "special studies" –"water resources", "water quality", "water ecology and rehabilitation", "groundwater", "small township sewage water treatment" and "Dagu river treatment". Due to page limits of this monograph,two of these – "water quality special study "and "water ecology and rehabilitation" are presented as examples that illustrate how the IWEMP special studies were implemented.

5.3.1 Special Study on Water Quality

5.3.1.1 Current Status of Water Environment Quality

(1) Rivers and Canals

Tianjin Municipality riverscape is comprised of 7 major drainage systems and includes 19 flood discharge rivers flowing (1,095 km), 79 irrigation drainage canals (1,363 km) and 2 sewage canals (105 km) Hai River basin includes totally. Water is seriously polluted; except for water transferred from Luan River and Yellow River for drinking water (Class II ~ III) water quality of in 2004 was , within the monitoring rivers course length of 1,638.7 km, the length of length of Classes II, III, IV, V and worse than Class V were 12.3%, 5.5%, 6.6%, 30.9% and 44.7 % for water of respectively.

The major drinking water source for Tianjin Municipality comes from Luan River-to-Tianjin Transfer project and Yellow River-to-Tianjin Transfer project. Water quality from Luan River is Class II~ III at each section. Yuqiao Reservoir, a major water supply reservoir, remains at Class IV~ V during the whole year due to influence of total nitrogen and total phosphorus; it has medium trophic state during the flood season. Water transferred from Yellow River has Class II ~ III at each section. Within the urban area water quality is mostly Class IV~ V with water quality better in the non-flood season in most of sections. The water ecological environment in "natural" rivers is vulnerable with low self-purification capacity. Due to influence of discharge of polluted water from upstream in flood season, the water quality dramatically decreases after rain and becomes Class 5 in parts of the river courses.

Pollution in canals that are primarily agricultural drainage is severe with water quality usually worse than Class V and accounts for more than 90% of the length of these watercourses. The major pollutants are NH_3-N, Permanganate Index, BOD and COD. There are two special drainage canals with total length at 116 km that play a special role in Hai River main course water quality protection; these discharge industrial and domestic wastewater, and also rainwater during the flood season from municipal and surrounding areas. For many years these two discharge canals have been seriously polluted and are worse than Class V. Water is murky, smelly, and has few living aquatic species.

(2) Marine Area

In 2004, the water quality in Tianjin marine area was mainly Class IV accounting for 57.1% of the monitored area. There are some differences in water quality between the wet/flood (summer) and dry (winter) seasons as follows: dry season> flood period> normal period; the major pollutants as lead and inorganic nitrogen. Seventy-one percent of the marine area is eutrophic with the highest eutrophic areas closer to shore. The eutrophication area in dry, flood, and normal periods is 14.3%, 71.4% and 85.7% respectively which is roughly equivalent to water quality status.

5.3.1.2 Tianjin Municipality Water Pollution Source Analysis

(1) Surface Water Pollution

In 2004, Tianjin Municipality river system received a total of 218,169.5 tons of COD, of which, 100,259.0 tons (46%) is incoming from outside the municipality's boundaries. Dagu drainage river, Ziyaxin River, Chaobaixin River, Ji Canal, Beijing Drainage River, and Qinglongwan River account for 75.2% of the total COD amount into the river and drainage system from the entire municipality. Of these the entire pollution load to Dagu drainage river was generated within the municipality whereas that for Ji

Canal is 71%, while the load in Ziyaxin River, Chaobaixin River, Beijing Discharge Canal,and Qinglongwan River was mainly from outside of the municipal boundaries. Ammonia (NH_3-N) amounted to 33,567 tons, with external influx of 18,180 tons (54.2%). The average NH_3-N load received by Ziyaxin River, Dagu Discharge Canal, Chaobaixin River, Beijing Discharge River, Beitang Discharge River, and Qinglongwan River was all above 2,000 tons, accounting for 82.3% of the total ammonia load.

(2) Pollutant Loads to Bohai Sea

Land-based pollution to the sea can be classified as: rivers into the sea, industrial enterprise sources, municipal and domestic source, integrated discharge sources and drainage canals to the sea, etc.. In 2004, the total water volume into sea from Tianjin Municipality was 1.235 billion m^3, with marine pollution load of COD and NH_3-N of 174,300 tons and 28.100 tons respectively. Most pollutants are discharged into sea from rivers, with COD and NH_3-N loads accounting for 65.3 % and 54.2 % of the total respectively. Detailed information on loads from various sources are noted in Table 5.2.

Table 5.2 Tianjin Municipality Pollutants Volume and Loads into the Bohai Sea, 2004

Number	Pollutants Source	Water Volume into Sea (100 million m^3)	Pollutants Load (Ton)	
			COD	NH_3-N
1	River	9.23	113,802	15,244.8
2	Industrial Enterprises	0.12	1,853	196.4
3	Municipal and Domestic Living	0.05	1,521	121.0
4	Integrated Outlets	0.27	3,661	140.1
5	Special Drainage River	2.69	53,455	12,445.7
6	**Total**	**12.35**	**174,292**	**28,148.0**

5.3.1.3 Tianjin Municipality Water Environment Issues

(1) Surface Water Pollution Problems

The major problem of Tianjin Municipality surface water pollution is water scarcity due to economic-social development and population increase in the upstream part of the basin. Water volume in the basin has been dramatically decreased with annual water volume flowing into sea from Tianjin Municipality decreasing from 14 billion m^3 in 1949 to less than 1 billion m^3 now. In some years there is zero water flow to the sea. Continuous drought and little precipitation in recent years create even more severe water resources conflicts. The limited water resources are now devoted to ensuring domestic living and industrial production; water for agriculture is no longer guaranteed, nor is water for maintaining basic river ecological functions.

(2) Pollution Source Problems

Industrial Pollution: In 2004, industrial sources COD and NH_3-N accounted for 26.4% and 23.7% respectively of the total pollutant discharge amount. Today, industrial pollution remains very serious. The problems can be categorized as follows:

- **Failure to comply with discharge standards:** Although over 99% of the industrial sewage discharge meet the standards in Tianjin in recent years, some enterprises still discharge COD that is above 500 mg/L (national sewage water integrated discharge Class III standard). For ammonia (NH_3-N) the national sewage water integrated standard limit value is 25 mg/L, however there is no specific requirement for NH_3-N within the Class III standard; however, the ammonia average discharge concentration for some major enterprises is above 25 mg/L.

- **Aging of industrial enterprises' water pollution treatment facilities, unstable operation and large pollutant discharges:** Investigations for 1,600+ enterprises in Tianjin Municipality indicated that existing treatment facilities are aging, unstable in operation or even failing in operation, leading some enterprises to exceed total volume requirements. In 2004, there were 73 enterprises with annual discharge amount of COD >100 tons and 13 >50 tons of ammonia; 184 enterprises unreasonably discharged sewage waters, 62 of them via ground penetration and/or evaporation.

- **Disparity between wastewater discharge standards, reduction targets and improvement of the water environment:** For a long period Tianjin followed the National GB 8978—1996 standard"Integrated Sewage Water Discharge Standard". The standard was promulgated in 1996; now, over one decade later, the technical and policy conditions that underlay that standard have changed dramatically, especially with rapid development of pollution prevention technologies, the standard value defined is too vague and the total pollutants discharge amount is not effectively controlled (the Tianjin Municipality sewage water integrated discharge standard has been promulgated but most enterprises did notfollow the new standard until 2010). A significant problem is that the national standard doesn't relate to surface water environment quality or the assimilation capacity that exists at specific locations.

A related problem is that current Sewage Water Integrated Discharge Standard doesn't regulate ammonia or total phosphorus discharged from factories and enterprises to sewage treatment plants. This leads excessive NH_3-N and phosphorus loads delivered to sewage treatment plants; ammonia and phosphorus reduction in treatment plants are not adequate therefore discharge of ammonia and total phosphorus from sewage treatment plants is above the standard and are the major factors for environmental deterioration in the Bohai off-shore area.

Domestic Wastewater: In 2004, township domestic source COD and NH_3-N discharge amount contributes to 53.1% and 54.4% of the total discharge amount respectively, with the advancement of urbanization, the domestic source pollution becomes more prominent. Township domestic source pollution problems mainly include following several aspects:

- **Large Wastewater Discharge:** By 2004, Tianjin Municipality has established 7 sewage treatment plants which could handle about 60.5% of total domestic wastewater. The rest was discharged directly without treatment.
- **High Concentration of Pollutants into the Sea and Poor Water Quality:** The rivers receiving township domestic sewage water are mainly agricultural water supply watercourses having water quality that was worse than Class V (COD 40 mg/L, NH_3-N 2.0 mg/L). Domestic sewage average concentration is far above the Surface Water Class V control standard, and part of the river pollutants concentration delivered into rivers is above agricultural irrigation water quality standard (standard value is 150 mg/L). Clearly, the surface water control standard could not be achieved.
- **Insufficient Treatment Capacity in Some Township Sewage Treatment Plants:** In 2004 the population served was 30.45 million people, mainly in 6 districts of the city, accounting for only 33.1% of the total municipal population. In some surrounding districts and counties of the municipality there is zero treatment and untreated wastewater is discharged directly to surface watercourses.
- **Wastewater and Rainfall Drainage Pipe Networks:** For the central urban area in 2004 the sewage network service area coverage rate is only 69.61%; the service area coverage rate for rainfall network is 59.09%.
- **Low Reclaimed Water Utilization Rate:** National standards require that the

reclaimed water utilization rate should be above 30% to 2010. However, there are only 2 reclamation water factories in Tianjin Municipality and the reclaimed rate is only 2%.

(3) Non-point Source (NPS) Pollution

- **Serious Surface Runoff Pollution:** Research in this Project shows that NPS pollution accounts for nearly 20% of the pollution from the whole municipality. NPS includes agricultural runoff and rainwater runoff from urban areas.
- **Low Centralized Livestock and Poultry Raising:** In 2004, centralized pig farm accounted for less than 50% of the total number of slaughtered animals; for poultry the rate is lower. Family farms and communal feedlots have no waste treatment; liquid manure mainly runs off into nearby water courses. Some manure is composted for field use, but much is lost in runoff during the rainy season.
- **Rural Domestic Living Wastewater is Un-Treated:** The rural population accounts for 40.5% of the total population in Tianjin Municipality. Most rural wastewater is discharged into pits, or directly to nearby watercourses without treatment.

5.3.1.4 Measures for Tianjin Municipal Water Environment Improvement

Two issues in particular need to be addressed. The first is to control incoming water quality and quantity at the municipality's boundaries; the second is to reduce and/or treat the discharge from the various pollution sources within the municipality. In this Project the targets by the end of 2010 are that major pollutants will be reduced by 10% compared with that of 2004; and in full compliance with the standards by 2020. As the Project has been implemented it is now estimated that by 2010, COD and NH_3-N discharge volume will be reduced by 10.1% compared to that of 2004 and will meet target requirements by 2020. However, in the absence of further engineering measures, surface water standards will not be achieved.

Reduction of pollution source discharge volume mainly includes engineering measure and management measures:

(1) Engineering Measures

- **Reduction Measure to 2010:** During the 11th "Five-Year" Plan period, Tianjin Municipality is to launch: 140 new industrial source treatment projects which will reduce COD by 43,900 tons, NH_3-N by 4,300 tons; 41 new domestic sewage treatment projects, and 45 reclaimed water projects, all of which would reduce

Chapter 5 INTEGRATED WATER RESOURCES AND ENVIRONMENT MANAGEMENT PLANS (IWEMPs)

COD by 67,800 tons and NH_3-N by 5,900 tons. It will launch 17 non-point source treatment projects which would reduce COD at 6,500 tons and NH_3-N at 900 tons.

- **Reduction Measures to 2020:** If the engineering measures during the 11th "Five-Year" Plan are fully implemented, it is predicated that COD and NH_3-N discharge by 2020 will be 246,000 tons and 30,000 tons. After reduction via projects to 2020, it is predicted that the COD and NH_3-N discharge will fall to 140,000 tons and 18,000 tons respectively. The discharges that are calculated to produce conforming water quality by 2020 are 131,000 tons of COD and 19,000 tons of NH_3-N. The actions to 2020 will be 48 new point source treatment projects that will mainly cover the Binhai New Area, four new districts, three counties, and two old districts. With these, it is predicted that COD could be reduced to 106,000 tons, which will exceed the target value by 25,000 tons; NH_3-N could be reduced to 13,000 tons which will exceed the target value by 6,000 tons. Except for Binhai New Area, the discharge in other areas cannot comply with standard, while the pollutants discharge amount of six districts within urban area, three counties and two districts is significantly different from target value, therefore, it is suggested either to enhance advanced treatment capacity or to establish new sewage treatment plants in the two regions mentioned above, to achieve the full compliance for Tianjin Municipality surface water.

(2) Management Measures

- **Industrial Restructuring with Target of Water Conservation and Discharge Reduction:** During the 10th "Five-Year" Plan Period, Tianjin Municipality successively reconstructed and adjusted state-owned large and medium enterprises. This included reducing those industries having low water consumption efficiency and high pollutants discharge volume. A number of enterprises with high consumption of water resources and energy, small scale, low technology, serious pollution and out-dated technology in textile and paper making industries successively were eliminated in Tianjin Municipality. Water consumption has first priority in industrial restructuring along with promoting high-quality, low consumption, and high added-value products. Fifteen categories of small polluting enterprises are forbidden.

- **Monitoring and Enforcement of Industrial Pollution Sources:** Monitoring needs to be strengthened at all levels of government monitoring. The Environment

Protection Department should implement integrated monitoring and management of major industrial pollution source treatment systems, organize treatment for major industrial pollution sources, and publicize industrial pollution source treatment information to society.

Major industrial pollution enterprises must receive their pollutants discharge permit certificate in accordance with law and discharge pollutants in accordance with the provisions of their permit. To accelerate industrial enterprises industry, it is necessary to continue to pursue structural adjustment of industry, to promote technology upgrading and clean production, and to promote conservation of water resources through water-use efficiency. The pollutants discharge reporting and registration scheme need to be enhanced. At technical level, it is necessary to enhance standardization of discharge outlets, to conduct online monitoring of major polluters, and to implement total volume (load) control for major pollutants such as COD and NH_3-N using actual assimilation values. In the areas that exceed assimilation capacity, the existing pollution source treatment degree must be enhanced and strict reduction measures must be taken to ensure that the pollutants discharge amount within the area will not exceed the total volume control indicator. For those areas that have assimilation capacity, the total volume control indicators should be reasonably allocated in accordance with the principles of openness and fairness, and pollutants discharge permit certificates issued based on these.

- **Greatly Enforce Clean Production Measures:** Clean production reduces pollutants within the manufacturing. Together with end-of-pipe control, clean production provides a balanced and cost-effective way of meeting discharge standards, especially as the cost of pollution control to meet increasingly stringent standards, solely with treatment, is becoming increasing expensive for industry.
- **Enhance maintenance and monitoring of municipal sewage systems:** Currently, most of the pumping stations in Tianjin Municipality were established in the 1950s or 1960s when the design standard was relatively low. Equipment is now outdated and the facilities now too small. Erosion by sewage water has caused the total capacity of pump stations to decrease; there are no valves for inflow control, and as the maintenance rate increases, there is serious leakage. There is now serious mixture of rainwater discharge and sewage water from separated drainage system. Although Tianjin Municipality adopted separated drainage system, due to

Chapter 5 INTEGRATED WATER RESOURCES AND ENVIRONMENT MANAGEMENT PLANS (IWEMPs)

lack of maintenance and other historical or artificial factors, mixture of rainwater and sewage is serious and causes over-load of treatment facilities. Investment for municipal pump stations should be improved, maintenance and reconstruction for pump stations should be carried out, and disused pump stations should be eliminated.

- **Promote Utilization of Reclaimed Water:** Integrated utilization of reclaimed water resources is one strategic measure for relieving chronic water shortage and for pollution prevention. It is also less costly than developing additional water sources. It is also more cost-effective than seawater purification; impurities contained in urban wastewater is less than 0.1% and can be removed by treatment. In comparison, seawater contains 3.5% salt and large amount of organic matter. Sewage reuse also reduced environment pollution, and has economic and social benefits.

5.3.2 Water Ecology and Rehabilitation

Human disturbance, decrease of freshwater amount into sea, and increase of pollutants discharge causes loss of ecological functions both in freshwater and in the marine environment. In freshwater, one of the most obvious losses is loss of wetlands due to unreasonable development, water scarcity, and eutrophication, and the drying up of rivers due to water scarcity, diversion and upstream water storage.

This study had the following objectives:

- Survey of water resources and water environment and selection of water ecology evolvement in 3 typical water bodies.
- Identify water ecology planning targets.
- Classify the water ecological zones and develop targets for different zones.
- Develop an indicator system of water ecology rehabilitation and protection.
- Determine the water resources and environment conditions for Tianjin Municipality water ecology rehabilitation, recovery and protection.
- Engineering projects and biological engineering projects to meet the water resources and environment requirements.
- Provide the 12th "Five-Year" Plan draft for Tianjin Municipality water ecology rehabilitation.

5.3.2.1 Water Ecology - Current Status and Trends

(1) Regional Water Resources Current Status and Trends

The following illustrate the ecological problems caused by water resources:

- Influx of water volume to the Bohai Seas has decreased dramatically over past decades to the point where there is virtually no input in some recent years (Table 5.3).
- Discontinuous of river flow and drying up of river course (Table 5.4).
- Gradual Shrinkage and Drying Up of Wetlands (Figure 5.1).
- Water Environment Quality Current Status and Tendency: In accordance with Tianjin Municipality Environment Quality Report (2001-2005), the water quality in major rivers in Tianjin Municipality is classified into Negative Class V or worse (Table 5.5).
- Changes in Marine Water Environment: Change in marine water environment can be classified into the following periods: dramatic change period of runoff into sea (1950s-1980s), massive land-based sewage discharge, massive marine area pollution period (middle of 1980s-middle of 1990s), massive discharge of nutrients, such as nitrogen and phosphorus (middle and late of 1990s), marine area high eutrophication and frequent red tide period (from 1990s until now). Sharp variation of runoff into sea brings problems like rise of salinity degree and off-shore fishery decline (Figure 5.2). Temporal changes in land-based sewage discharge and marine area pollution are indicated in Figure 5.3 and Figure 5.4 After the 1990s, increases in primary productivity, illustrated typically by increase of Chlorophyll, causes frequent red tide in off-shore areas (Figure 5.4).

Table 5.3 Variation of runoff to Bohai Sea in Tianjin Municipality (100 Million m^3)

Runoff to Bohai Sea Time Period	Annual Average Runoff (Calendar Year)	Flood Season	Non-flood Season	Percentage of Flood Season (Calendar Year)/%
1950-1959	144.3	89.3	55.0	61.9
1960-1969	81.7	49.8	31.9	61.0
1970-1979	33.1	27.7	5.4	83.7
1980-1989	9.8	8.5	1.3	86.3
1990-1999	7.7	6.8	0.9	88.3
2000-2005	3.8	3.6	0.2	94.7

Chapter 5 INTEGRATED WATER RESOURCES AND ENVIRONMENT MANAGEMENT PLANS (IWEMPs)

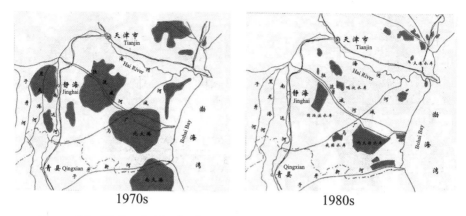

1970s   1980s

Figure 5.1 Wetlands Distribution in Middle and Mouth of Tianjin Municipality

Table 5.4 Discontinuous flow and drying up of major river courses in Hai River basin

Name	Section	Section Length/km	Average Annual River Course Drying Up Period/Day				
			1960s	1970s	1980s	1990s	2000
JI Canal	Jiuwang Village-Xinfangchao Dam	189	2	33	115	257	365
Chaobai River	Su Village-Ningchegu	140	4	142	184	197	300
Bei Canal	Tong County-Qujiadian	129	8	118	126	202	310
Yongding River	Lugou Bridge-Haikou	199	198	312	362	365	366
Main Course of Hai River	Er Dam-Haihe Dam	73	0	0	0	0	0
Ziya River	Xian County-Diliupu	147	84	280	349	328	366
Nan Canal	Sinv Temple-Diliupu	306	32	207	320	341	366
Name	Section	Section Length/km	Average Annual Discontinues Flow Period/Day				
			1960s	1970s	1980s	1990s	2000
JI Canal	Jiuwang Village-Xinfangchao Dam	189	33	41	300	312	365
Chaobai River	Su Village-Ningchegu	140	45	194	319	197	366
Bei Canal	Tong County-Qujiadian	129	99	270	242	358	340

Name	Section	Section Length/km	Average Annual Discontinues Flow Period/Day				
			1960s	1970s	1980s	1990s	2000
Main Course of Hai River	Er Dam-Haihe Dam	73	129	265	298	254	332
Ziya River	Xian County-Diliupu	147	124	295	354	328	366
Nan Canal	Sinv Temple-Diliupu	306	53	175	302	341	366

Table 5.5 Water quality in the 17 Class One rivers of Tianjin Municipality (2001-2005)

Number	River Names	2001	2002	2003	2004	2005
1	JI Canal	<Negative Class V	<Negative Class V	<Negative Class V	<Negative Class V	Negative Class V
2	Gou River	Negative Class V	<Negative Class V	Negative Class V	Negative Class V	Negative Class V
3	Huanxiang River	<Negative Class V	<Negative Class V	<Negative Class V	<Negative Class V	<Negative Class V
4	Gou-Chao River Water Transfer	<Negative Class V	<Negative Class V	<Negative Class V	<Negative Class V	<Negative Class V
5	Chaobaixin River	<Negative Class V	<Negative Class V	<Negative Class V	<Negative Class V	<Negative Class V
6	Qinglongwan River	<Negative Class V	<Negative Class V	<Negative Class V	<Negative Class V	Negative Class V
7	Bei Canal	<Negative Class V	<Negative Class V	Negative Class V	Negative Class V	Negative Class V
8	Beijing Discharge Canal	Dry	<Negative Class V	<Negative Class V	<Negative Class V	<Negative Class V
9	Yongding River	<Negative Class V	<Negative Class V	Dry	Dry	Dry
10	Yongdingxin River	<Negative Class V	<Negative Class V	<Negative Class V	<Negative Class V	<Negative Class V
11	Jinzhong River	<Negative Class V	<Negative Class V	<Negative Class V	<Negative Class V	Negative Class V
12	Ziya River	<Negative Class V	<Negative Class V	<Negative Class V	Class IV	Negative Class V
13	Duliujian River	<Negative Class V	<Negative Class V	<Negative Class V	<Negative Class V	<Negative Class V

Chapter 5 INTEGRATED WATER RESOURCES AND ENVIRONMENT MANAGEMENT PLANS (IWEMPs)

Number	River Names	2001	2002	2003	2004	2005
14	Daqing River	<Negative Class V	<Negative Class V	Negative Class V	Negative Class V	Negative Class V
15	Nan Canal	Dry	Class IV	Class IV	Class IV	Class IV
16	Machangjian River	<Negative Class V	<Negative Class V	Dry	<Negative Class V	<Negative Class V
17	Ziyaxin River	<Negative Class V	<Negative Class V	<Negative Class V	<Negative Class V	<Negative Class V

Note: <Negative Class V mean worse than Class V (the worst category of water quality).

(2) Typical Water Body Water Ecological Current Status and Evolvement

Drinking water sources, scenery water body, and the Dagu river estuary marine area are selected as typical water bodies. Current status investigation was conducted during September to October in 2005, in combination of historical monitoring data. The following three examples demonstrate the impacts:

- Yuqiao Reservoir, a major drinking water source for Tianjin, is increasingly eutrophied by oxygen-demanding organic matter, and nitrogenous and phosphorus organic matter – both arising mainly from sewage from Zunhuan County urban areas in Hebei Province and the LI river system.
- The Hai River main course urban area section (a "scenery" river section) is seriously polluted from urban sewage water discharged by combined sewage systems (rain+sewage) during flood season and from phosphorus release from riverbeds during non-flood season.
- The Class IV area of the Dagu River/Canal estuary is expanding and marine ecology is threatened both by pollution and eutrophication and is deteriorating. The major reason is the (temporal and spatial) lack of freshwater influx into the sea.

Recommendations to Improve the Tianjin Water Ecology Situation

(i) Water resources requirements for freshwater ecological rehabilitation

Table 5.6 provides the short-term, middle-term and long-term water ecology water demand amount with minimum, suitable and optimum water ecological water demand volumes.

(ii) Freshwater volume to the sea

For the period 1980 to 1989, 30.47% is used as the minimum seawater/freshwater salinity benchmark. In contrast, using average salinity during April to

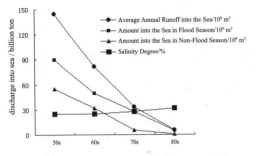
Off-shore Marine Area Salinity Degree Rise

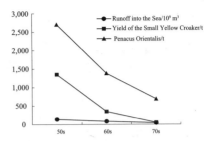
Off-shore Fishery Decline

Figure 5.2   Impacts from sharp variations of runoff to the Bohai Sea

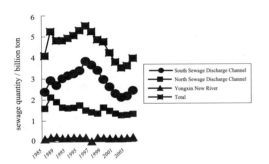
Sewage Water Volume Variation from Mid-1980s till Now

COD Amount into Sea Variations from Mid-1980s till Now

Figure 5.3a   Temporal Trends in Pollutant Discharge into the Bohai Sea

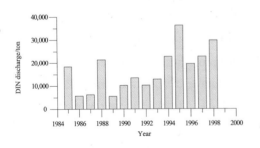
Amount of Total Nitrogen into Sea Variation Tendency

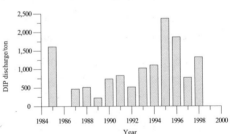
Amount of Total Phosphorus into Sea Variation Tendency

Figure 5.3b   Temporal Trends in Pollutant Discharge into the Bohai Sea

Chapter 5 INTEGRATED WATER RESOURCES AND ENVIRONMENT MANAGEMENT PLANS (IWEMPs)

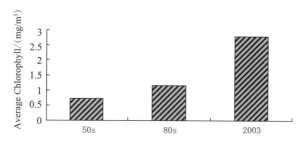

Figure 5.4 Average Off-shore Area Surface Chlorophyll

November from 1955 to 1982, 28.23% would be the salinity benchmark. To meet the first benchmark, the minimum freshwater amount into sea should be 1.55 billion m^3 annually. To meet the latter benchmark, volume of inflow to the sea should be 1.818 billion m^3 annually.

Table 5.6 Major Water Body Ecological Water Demand Volume (100 million m^3)

Project		Short Term	Middle Term	Long Term
Urban Rivers and Lakes		2.94	3.99	4.06
Wetlands		3.74	4.59	5.94
Riverbeds (Loss)		4.43	6.22	16.55
		11.09	14.80	26.55
Drinking Water System (Reservoirs)		1.05	1.05	1.05
Base Flow of Central Rivers	Three Estuaries 1950-2000	5.06	17.99	28.10
	Six Cross-sections 1980-2004	3.24	—	
Total	Calculated by three estuaries	18.01	33.84	55.70
	Calculated by six cross-sections	16.19		

(iii) Water Resources Source Analysis

Considering the entire Municipality, water volume from the Luan River Transfer Project is 1 billion m^3/year; it is planned to transfer 1 billion m^3/year from Jiangzhong Route. The local river flow is 0.05 billion m^3/ year, and the (treated) sewage water resource amount is 0.95 billion m^3/ year. The total estimated water that is, or could be made available is 4 billion m^3/year after deducting domestic and production water consumption. This would meet the ecological water resources requirements with reasonable planning.

(iv) Water Environment Pollution Treatment

Total pollutants reduction of COD of 556,000 tons /year will meet the Class V water quality target, and the drinking water source target of Class III water quality. These would meet the minimum standard required for water ecology rehabilitation.

(v) Water Ecological Function Zoning

In combination with water function zoning and water environment function zoning, and coordinating with flood prevention planning, Tianjin Municipality could be divided into five water ecological function zones as shown in Figure 5.5. The research establishes water quality, water quantity and water ecology indicators system on the basis of specific rehabilitation tasks of five major function zones.The requirements for these are:

- Hai River Main Stream-Municipal River and Lakes Hydrology Water Ecological Function Zones rehabilitate and reduce smelly, lifeless water body, to meet the requirements of Negative Class V water body.

- Rehabilitation of Qilihai-Dahuangpu wetland water function zone in the middle area and Tuanpowa-Beidagang wetland water function zone in the southern area. Water ecology system environment conditions, wetland purification and adjustment functions, meet the water quality and quantity requirement of freshwater into sea.

- Off-shore marine water ecology function zone: prevent continuous deterioration of water ecology, ensure quality and quantity of fresh water into sea, implement breeding and releasing project.

- North drinking water source water ecology protection area: protect water ecology system, to ensure the water quality of water body is stably above

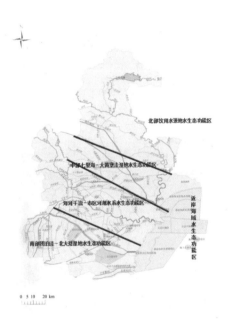

Figure 5.5 Ecological zones proposes for Tianjin Municipality

surface water Class III quality in the long-term.
(vi) Water ecological guarantee system

The water ecology rehabilitation implementation guarantee system mainly includes the necessary organizations, institutions, laws and policies and planning systems.

5.4 TYPICAL COUNTYLEVEL IWEMPLANS

5.4.1 Pinggu District of Beijing Municipality

Pinggu district of Beijing Municipality is located in the upstream of the Su canal river system of Hai River Basin, in the outer suburbs of east Beijing. The main river, Xun River, has a 1,692 km² watershed area. Its source is Xinglong County of Hebei Province, flowing through Ji County of Tianjin and then into the Haizi reservoir in east Pinggu district. The river passes through 10 townships en route to Sanhe city of Hebei, Pinggu district is comprised of two landform units — a south west plain region of 402 km² and mountainous area of 540 km² to the north, east, and south. Pinggu district has one of the highest rainfall of Beijing. From 1980-2005 the Pinggu district average annual rainfall was 614.4 mm with little difference between mountain and plains areas. Average annual surface water resources (useful) amount is 71.81 million m³; average years groundwater exploited amount is 163.40 million m³. River water quality in mountain areas is good (Class II) but water pollution is serious around Pinggu district and downstream water quality classes are Class V. There are 9 reservoirs in Pinggu district with total storage capacity of 148 million m³. In 2004 Tianjin constructed a diversion from the Xun River so that flow in Pinggu has decreased greatly. Now Pinggu district uses water mostly from groundwater with little surface water use. In 2005 Pinggu district began supplying water to Beijing with an annual volume of 80 million m³.

In 2005 Pinggu district established the district Water Bureau to manage township water supply, drainage and sewage treatment. However, the district Environmental Protection Bureau remains in charge of household sewage, etc.. Currently, the main water resources problems in Pinggu district are: decreasing flows into the district, and increasing flows out of the district; the results is that there has been a major decrease both in surface and groundwater that now creates water shortages and environmental deterioration.

5.4.1.1 Objectives for the Pinggu District IWEMP

World Bank experts introduced Pinggu district to IWEMP planning based on ET management to reduce non-beneficial ET water consumption, with saved water used to restore surface and groundwater resources. Also, by strictly controlling pollution discharges, bring water environment function zone up to required standards. The overarching objective was to make Pinggu district water resources sustainable, to rehabilitate water function zones and water environment function zones, to improve the social economy, and to promote sustainable water resources and water environment development.

Specific objectives of the IWEMP are:

- Water resources sustainability (water amount objective):By 2010, reduced groundwater overdraft by 10%, and to achieve a balance between groundwater exploitation and recharge by 2020.
- Maintain healthy ecological environment (water quality objective):By 2010, reduce wastewater discharges to rivers by at least 10%, and attaining water function zone standards by 2020.

5.4.1.2 Planning Approach

The IWEMP was based on the concept of controlling consumption of ET, and an integrated approach to water resources and environment management. A water balance approach was used which included determining actual and target ET withdrawal control, consumption control, reuse and release. Pollution management was developed through determining total current pollution load and measures to achieve a target pollution discharge volume.

(1) Integrated ET Assessment

ET was assessed for a variety of land classes (Table 5.7). The objective was to understand the role of anthropogenic activities in increasing ET consumption (loss) and therefore increasing the un-sustainability of water resources. Integrated ET is composed of natural ET and anthropogenic ET. The IMEMP would propose management measures to reduce ET in different classes by reducing anthropogenic ET such as in irrigation. Mitigation measures would include agricultural measures such land leveling, mulching covers, greenhouse construction, planting structure adjustment, fallow, rainfall agriculture, etc..

(2) Modelled Scenarios

Scenario analysis involved the "3S" technology (GIS,RS [remotes sensing] and GPS), distributed basin hydrology model (SWAT), soil water balance model (AquaCrop), river water quality simulation tool (WQS model), together with a structured Pinggu District water resources and water environment planning model. For water resources consumption balance, the analysis used the SWAT and AquaCrop models, conducted basin consumption balance analysis at basin and field levels according to the framework noted in Figure 5.6.

For water environment load capacity analysis, the water quality simulation tool (WQS Tool) simulated point source and nonpoint source pollution influence on regional river quality under specific flow and wastewater discharge conditions in order to evaluate water environment load capacity in different water function zones in Pinggu District.

Pinggu District water resources and water environment model was a simple lumped model, covered all human impacted links such as supply, allocation, consumption, drainage, pollution control, reuse water, etc.. Target ET was factored into the modelled scenarios in order to calculate the appropriate water allocation for various water resources purposes, including permitted water amounts for withdrawal, consumption, reuse and release.

Table 5.7　Regional Integrated ET Land Classification

ET land classes	Regional integrated ET classify		
	Ecological water use (ECO)	Agricultural water use (AGR)	Urban and rural water land use (URB)
Land	Nature mountain forest, reservoir, pond, lake, road, bare land	Farmland, fruiter, fishpond	Urban and rural residential areas, industrial areas, cultivation areas
ET composition	Only nature ET	Nature ET + Manpower ET	Natural ET + Anthropogenic ET
Influence ET Factors	Climate condition, underlying surface(water area, vegetation cover, etc.)	Climate condition, underlying surface Crops Structure and fishpond area, Planting form, agriculture management, irrigation method, water saving management, etc.	Climate condition, underlying surface harden degree, supply and discharge method, industrial structure, Industrial water technology, domestic water saving method, etc.

ET land classes	Regional integrated ET classify		
	Ecological water use (ECO)	Agricultural water use (AGR)	Urban and rural water land use (URB)
Water circulation way	Natural water circulation	withdrawal water participated in Nature water circulation	withdrawal water formed human water circulation, separate with Natural water circulation
Anthropogenic	No	withdrawal water through water area, crop(plant), soil produce ET	withdrawal water produce consume water in domestic, production

5.4.1.3 Pinggu District IWEMP Process

There were six steps in Pinggu District IWEMP Process:

- Current data baseline survey: Based on water resources, water environment, social economy etc., a KM data platform was established for Pinggu. Utilized 3S technology and Pinggu 1m image resolution aerial photo remote sensing map processed water and soil information baseline survey to Pinggu District land use. A land use survey at a scale of 1 ∶ 10,000 provided necessary data for determining consumption balance using land data, a basin distributed hydrology model, and ET management requirements.
- Develop current water use using SWAT, AquaCrop model etc., and pollution load capacity simulation using the water quality simulation tool.
- Established target ET and target pollution levels: Established the target ET for the plains (agricultural area) using average precipitation values. The river water quality simulation model (WQS model) allowed the determination of allowable wastewater discharge target.
- Select management scheme scenarios: Selected/adjusted management measures using AquaCrop.
- Management scheme recommendation: Final implementation management scheme that can achieve the various targets is recommended in the IWEMP.
- Control Indicators: Proposed water resources development and utilization control indicators, water using efficiency control indicators, and water function zone restricted pollution indicators.

5.4.1.4 Plan Outcomes and Implementation

The planned outcomes contained in the Pinggu District IWEMP are shown in Table 5.8 together with the indicators and targets for 2010 and 2020. These targets will be achieved using the following implementation scheme.

(1) Decrease ET Loss in agriculture

- Adjust agriculture land and crop structure: Decrease 50% for fishpond, decrease winter wheat area – decrease 7,500 mu for double crop (multiple crop), decrease 16,500 mu for open ground vegetable, increase green house rate by at least 70%.
- Optimize irrigation system: Popularize drip irrigation technology; optimize winter wheat irrigation system according to soil moisture with deficit irrigation; reduce irrigation time; optimize vegetable irrigation system; adopt micro

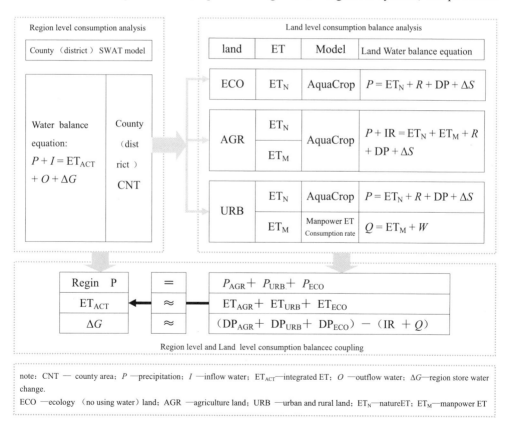

Figure 5.6 Region and Field Level Consumption Balance Coupling

and drip irrigation technology in greenhouse agriculture; increase use water efficiency.
- Agricultural water saving measures: Improve agriculture water saving construction, during 12th "Five-Year" Plan by increase of 35,000 mu, transform and renovate water saving areas by 75,000 mu, water saving engineering match rate reaches 100%, strengthen agriculture wells metrology management, practice "one well-one meter-one household-one card-one number", during 12th "Five-Year" Plan install 1,500 water meters in 10 townships, practice long distance metrology control in agriculture wells.

(2) Decrease water consumption in urban and rural areas
- Adjust industry structure, develop high production value, low consumption water industry such as electronics and communication equipment manufacturing.
- Integrated water supply system in new city: when constructing new water plant implement water resources transfer engineering; concentrate supply water substitute for well supply; practice unified management; decrease pipe network leak rate to under 10%.
- Construct urban and rural integrated water supply network system in Pinggu District, by linking the main pipe with every integrated supply water unit by 2020; form unified supply water network as different water resources in plain areas, realize unit scheduling to surface water plant and groundwater plant; increase supply water efficiency and guarantee rate.
- Construct separate rainwater and wastewater systems; decrease evaporation loss produced by discharges.

(3) Increase water use efficiency
- Strengthen management of industrial enterprises; set a realistic industry water price; promote industry water saving, improve industry enterprise water saving consciousness.
- Practice regular water balance tests in various enterprises; improve water using technology, increase industry value per unit of consumed water.
- Increase water saving equipment popularizing rate, especially to rural area, during 12th "Five-Year" Plan; new city water saving equipment popularizing rate reaches 100%, rural area reaches 70%.

Chapter 5 INTEGRATED WATER RESOURCES AND ENVIRONMENT MANAGEMENT PLANS (IWEMPs)

- Urban and rural supply water practice metering information, practice using water amount progression increasing price regulation, control water use at household level, encourage household water saving.
- In industry and institutions, establish participation mechanism involving society to mobilize water saving enthusiasm to every water using unit.

(4) Increase water reuse measures

- Study increasing reuse water possibility in industry, and increase water reuse rate technology.
- During the 12th "Five-Year" Plan, construct water reuse plant in new city and focal township; greatly increase water reuse production for use both in domestic use and to improve river water quality.

(5) Water environment rehabilitation

- Practice "Haiguiji [2002] 68 meeting summary; establish long-term water allocation mechanism to Tianjin Xun river, with 70% guarantee rate; Tianjin Yangzhuang should release 13.86 million m^3 to Haizi reservoir.
- Construct west river pollution treatment plant in new city; by 2015 urban pollution treatment rate will reach 93%, and in 2020 will reach 100%. Construct urban and rural intensification three level pollution treatment systems; construct concentrated pollution treatment plants in 10 townships by 2015; township and village pollution treatment rate reaches 71%. Total township concentrated pollution treatment plant will be constructed before 2020; township and village pollution treatment rate will reach 90%.
- Decrease nonpoint source pollution; all small basins will construct to ecological cleaning small basins before 2015, construction rate will be 100%.
- Through river ecology construction and new city water system construction, new city river section water quality main indicators reaches water function zone standards.

Table 5.8 Pinggu District IWEMP Targets and Indicators

Classify	Indicators	Unit	Current 2005	2015	2020
Social economy	District population	10,000 person	39.5	45.4	61.4
	Urbanization rate	%	39.7	60.4	72.6
	District GDP	100 million CNY	57	214	390
	District industry increase value	100 million CNY	24.3	96	172
	Unit sheep	10,000	45	55	65
	Fishpond area	mu	13,500	6,750	6,750
	Total irrigation area(farmland, fruiter, vegetable)	mu	341,250		324,750
	Food production	10,000 ton	4.25		4.9
District total water consumption (integrate ET)	District target ET	mm			533.3
	District (942 km^2) plan integrate ET:	mm	544.7	537.1	532.9
		10,000 m^3	51,311	50,594	50,196
	Ecology land (551.1 km^2)	10,000 m^3	27,042	27,042	27,042
	Urban and rural land (100.7 km^2)	10,000 m^3	6,137	6,113	6,189
	Agriculture land (290.2 km^2)	10,000 m^3	18,132	17,457	16,965
Using water total amount control	District social economy water using amount	10,000 m^3	14,295	12,885	13,165
	Surface water withdrawal amount	10,000 m^3	0	1,150	1,150
	Groundwater withdrawal water restrict	10,000 m^3	14,295	10,835	10,215
	Urban and rural land (township +countryside)	10,000 m^3	4,423	3,860	4,240
	Agriculture land (irrigation +fishpond)	10,000 m^3	9,872	6,975	5,975
Groundwater transfer from outside	Emergency sources supply water to Beijing	10,000 m^3	5,100	8,000	0

Classify	Indicators		Unit	Current 2005	2015	2020
Anthropogenic ET consumption control	District anthropogenic ET:		10,000 m³	4,671	3,935	3,650
	Urban and rural land (township +countryside)		10,000 m³	1,726	1,665	1,777
	Agriculture land (irrigation +fishpond)		10,000 m³	2,945	2,270	1,873
	Anthropogenic ET consumption rate		%	32.7	30.5	27.7
Using water efficiency control	District using water	Per capita using water	Liter / person / day	992	778	587
		10,000 yuan GDP using water	m³/10,000 CNY	251	60.2	33.8
	District consumption water (manpower ET)	Per capita consumption water	Liter / person / day	323	238	162
		10,000 yuan GDP consumption water	m³/10,000 CNY	81.9	18.4	9.4
Reuse water and release water amount	Reuse water production		10,000 m³	0	2,430	3,400
	Reuse water for industry and domestic		10,000 m³	0	900	1,800
	Urban and rural land release water (include reuse water)		10,000 m³	2,697	2,645	2,913
Permit pollution control	COD pollution restrict		ton			2,300
	NH₃-N pollution restrict		ton			220
	River water quality around new city		Class	< V	V	V

(6) Water resources and water environment management measures

- Strengthen basic level water station capacity; complete a three level water management system. Strengthen water saving, supply water, pollution treatment, flood control, water resources protection, etc. management function in basic level water station. Improve and effective management at the village level using farmer water users associations; improve 900 water managers team to be in charge of village water management.
- Establish agricultural ET management center; practice ET monitoring and

evaluation system. The agricultural ET monitoring and evaluation system is shown in Figure 5.7.
- After the South to North water diversion project is completed, water resources will be managed by district government; there is need to establish a unified water supply management institution, construct water supply network information system, covering surface and groundwater. Establish water saving management assessment regulation, detail water saving statistical analysis, and practice integrated water assessment for water saving achievements.
- Establish Pinggu District river canal water network management institution, practice pollution abatement. After construct east and west water network and new city water system, establish Pinggu District river canal network management institution participated by water bureau and environment protection bureau, and also establish Pinggu District water network unified scheduling information platform, in charge of scientific scheduling for inflow water, reservoir canal water, reuse water to river, in charge of pollution discharge, water function zone water quality monitor management.
- Establish water reuse management institution and reused water policy mechanism involving environment, afforestation, domestic toilets, dust reduction, and for replenishing surface and ground water.

(7) 12th "Five-Year" plan indicators and emergency measures

Since the South to North water diversion project is delayed, Pinggu District groundwater capacity is not good. In order to not negatively affect economic development

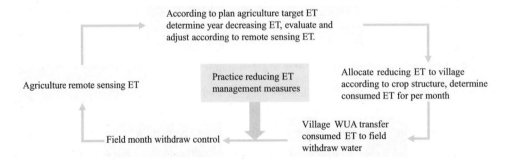

Figure 5.7　Pinggu District Agriculture ET Management Monitoring and Evaluation System

during the 12th "Five-Year" Plan, emergency measures are identified that are necessary to ensure that there is no un-necessary overexploitation of groundwater. During 13th "Five-Year" south to north water will deliver to Beijing, emergency water sources will not supply water to out borders, after 2015 groundwater will be relatively plentiful, can reach groundwater exploitation and replenishment balance.

5.4.1.5 Summary and Experience

The IWEM Plan demonstrates integrated water resources and water environment management planning involving the water and environment sectors. Water quantity targets are based on ET evaluation with rational development of implementation strategies based on sound technical understanding. The Plan contains quantitative targets and measurable indicators to evaluate Plan implementation through to 2020.

5.4.2 Xinxiang County of Henan Province

Xinxiang County is an example of a small city that is typical of the Hai River basin. The county is now water scarce, in part because of a larger papermaking factory that uses huge amounts of water and discharges large wastewater loads to the environment. The objective was to develop and IWEM Plan using ET monitoring technology to develop a rational approach to water use management and water use reduction both in rural and urban areas. Also, the Plan was to develop pollution abatement strategy in order to achieve water quality targets in surface water of the county.

Main outcomes and experiences in Xinxiang County

(1)"Top down and bottom up" collaborative mechanism

Xinxiang county developed "top down and bottom up" mechanisms of collaboration and planning. "Top down" reflected requirements imposed by higher level entities such as the CPMO, the Zhangweinan sub-basin office level, the various ministries involved. And guidance provided downwards by the county to township and local levels. "Bottom up" refers to inputs from the township, village and water user levels that was used to develop appropriate policies and practices that were consistent with local level needs and abilities.

(2) Multi-sectoral collaborative mechanism for, mainly, water resources and water environment sectors

According to the design requirement of this GEF project, Xinxiang county realized coordination and collaboration across sectors, organized by county environment and water sectors, with close cooperation from agriculture, construction, and land resources sectors. This allowed holistic compilation of the IWEM Plan. To ensure project implementation, a project leading group was established in Xinxiang county and under it was a GEF project management office and joint expert panel. The Project Office developed suitable working arrangements, established consultative mechanism, invited representatives of the government, and of other stakeholders including farmer water using associations and local communities, of vulnerable groups, and ensured information dissemination and communication. The Project Office ensured cooperation among project and stakeholders, held consulting and training seminar, asked for views, comments and recommendations in various levels, maintained meeting records, and drew up the project action plan to ensure suitable project implementation.

(3) Positively participated in technical training and workshop, promoted technical communication and cooperation

Xinxiang county participated in at least 10 training sessions and workshops to develop that compilation required for the IWEMP. These sessions included SWAT model development, remote sensing monitoring of ET, outcomes analysis, project monitoring and evaluation that was organized by the CPMO. Through training and workshopsthe project concept was promoted at all levels.

(4) Main technical outcomes

- The Xinxiang County integrated water resources and water environment management information system was developed using the KM platform developed by the larger GEF Project. In addition to standard software and GIS, the information system contains baseline data, models such as SWAT land use map, soil map, weather, hydrology, agriculture management, water using management and pollution data. This platform allowed development of an integrated system of data management and data sharing amongst the collaborating institutions.

- Using ARCGIS as a platform, and accessing remote sensing technology, the Xinxiang County dualistic water cycling system based on SWAT-MODFLOW artificial variables coupling. Through water resources evaluation, evapotranspiration simulation, groundwater prediction, point source and

Chapter 5 INTEGRATED WATER RESOURCES AND ENVIRONMENT MANAGEMENT PLANS (IWEMPs)

nonpoint source pollution simulation, using dualistic water cycling system, water cycling and water quality targets, determined the basis for integrating water resources and water environment management schemes.

- Based on water balance and water environment load capacity analysis, determined key indicators at the county level for the IWEM Plan.
- Recommended technological and management systems for water saving and pollution control. According to Xinxiang County situation, established different scenarios for water saving, for pollution control, and for management measures, and recommended optimum schemes for each of these.
- Introduced the "real water savings" concept to water resources management through total ET control, increased ET efficiency, and decreased ET loss.

(5) Xinxiang County water pollution control demonstration project

The demonstration focused on the Henan Longquan Industry Ltd. which makes paper and has very high pollution load to the environment. The Project developed control and investment options that will have the following results: a daily saving of water resources of 16-18 thousand tons, and wastewater discharge will decrease markedly.

5.4.3 Guantao County of Hebei Province

Guantao county is located in the southern plain area of Hebei Province, upstream of Zhangweinan canal of Hai River Basin. The total area is 456 km^2. It is warm temperate and semi-humid region, with average rainfall of 531.6 mm. Guantao county is mainly a floodplain with soil layers of sand and clay. Shallow groundwater depths are 50-80 m with single well yields of 30-40 m^3/h; middle groundwater layers are 80-250 m deep and are saline with mineralization of 2-10 g/L. Deep layers (>250 m) are deep fresh water, sand layers are 40-60 m, with single well yields of 60-80 m^3/h, The deep aquifer is recharged from upstream recharge areas but recharge water is small therefore groundwater extraction should be strictly controlled.

Zhang river originates in Shanxi province, whereas the Wei river originates from the Jiaozuo region of Henan province. These two rivers join as Wei canal, part of the Hai River Basin system. Average runoff is 980 million m^3/year with 80% of inflow water concentrated in the rainy (flood) season in June to September. In other seasons rainfall runoff are very little. River water is very polluted. Guantao county suffers drought in the spring when water is required for wheat. In summer, during the flood season, river water is plentiful but is not needed for irrigation. Therefore, the Weixi canal stores water to recharge groundwater.

Guantao County has 8 townships with a total population of 288,000. Agriculture accounts for 265,000 persons. Cultivated lands are 430,000 mu, and per capita cultivated land is only 1.6 mu for the agricultural populations. Irrigation land is 405,000 mu which is 94% of the cultivated lands. Crops are mainly wheat, corn, cotton, vegetable and oil crops. Since the 1990's, there has been much crop adjustment; now there are efficient farming regions, vegetable greenhouses, orchards, open air green vegetables, etc.. Chicken raising has become very important with production value increasing to 48% from 5.6% in the early 1980's.

The region produces internal surface runoff of 56.994 million m^3; irrigation diverted from Wei canal is 7.333 million m^3; average annual groundwater use is 56.921 million m^3. Average annual water utilization total is 66.794 million m^3. From 2014, Guantao county is to receive 5.6 million m^3 from the South-North water diversion project. At that time average annual utilization will be 72.394 million m^3.

Guantao county is short of surface water during the growing season and has limited groundwater, therefore, the county suffers acute water shortage. On the other hand, water use efficiency is very low, and water quality is degraded from a variety of pollution sources. As a consequence there is overexploitation of groundwater resulting in groundwater table levels decreasing every year. Groundwater depth decreased from 14.9 m in 1994 to 22 m in 2004 –a decrease of 7.1 m in 11 years.

5.4.3.1 Objectives of the IWEM Plan

(1) Overall objectives

The overall objective is to achieve a balanced water use and water pollution control strategy based on modern scientific methods, and integrated and modern water management methods. The long-term objective is to balance water use with water withdrawal in order to stabilize the groundwater table. Also, the overall objective is to bring stakeholder participation into the management decision-making process and the bring farmers to understand that they must be part of the solution.

(2) Specific objectives

- Management objectives: establishes and perfects integrated water resources and

Chapter 5   INTEGRATED WATER RESOURCES AND ENVIRONMENT MANAGEMENT PLANS (IWEMPs)

water environment management system and mechanism, draws up and practices relative policies and laws, improves management capacity in integrated water resources and water environment management.

- Water saving objectives: control water use with target ET values. By 2010 ET should decrease by 10% compared with baseline. By 2020 there should be, on average, a further decrease in ET loss of 10%. In 2020 there should be a water resources supply and consumption balance. This accommodates variation in ET and water supply and use between wet and dry years.

- Pollution objectives: Improve pollution treatment so that by 2010 there is a 10% decrease compared with the baseline year. By 2020 water quality should attain surface water quality standards.

- Groundwater stabilization: Decrease groundwater overexploitation by 10% by 2010, with zero overexploitation by 2020.

5.4.3.2   Summary of Plan Methodology

Guantao County IWEMP compilation is supported by KM platform and tools, and remote sensing ET monitoring technology. Through project implementation, establishes integrated water resources and water environment management system and mechanism, draws up and practices relevant policies and laws, improves integrated water resources and water environment management capability. The framework for the Guantao IWMEP is illustrated in Figure 5.8.

5.4.3.3   Main Outcomes of Plan

(1) ET reduction and water resources supply linkage

Current water supply is 258.188 million m^3 whereas consumption is 277.613 million m^3 leaving a negative amount of 19.42 million m^3. This converts to 42.6 mm of ET. Therefore, decreasing ET by 42.6 mm will meet the water deficit. A further decrease of ET by 30.4 mm decreases water consumption further by 13.862 million m^3.

(2) Water resources supply and consumption

Using ET targets as a basis for making regional calculations, and knowing that water supply includes: rainfall, Weixi main canal storage water, and Wei canal leakage water (to groundwater), the IWEM Plan allocates Weixi main canal storage for townships' cultivated areas prorated according to each township cultivated areas; Wei canal leakage water is allocated to the adjacent four townships (Guantao township,

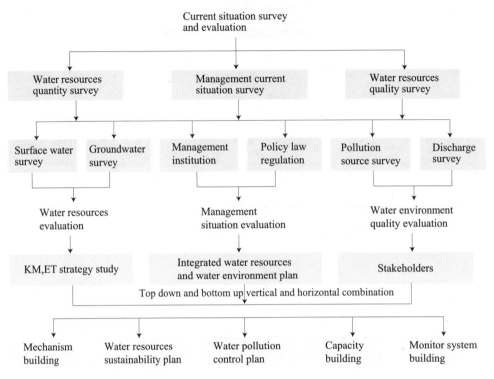

Figure 5.8 Guantao County Integrated IWEM Planning Framework

Weisengzhai township, Wangqiao township and Xucun township). The targets are shown in Table 5.9.

(3) Groundwater

According to the 1980-2005 period the Plan calculates water supply water, target ET, and surplus and shortage of water under various scenarios and conditions. The Plan takes into consideration groundwater recharge volumes under differing conditions and calculates the allowable groundwater exploitation. The objective is to achieve groundwater dynamic balance in average rainfall years. The groundwater targets under different hydrological conditions are shown in Table 5.10.

(4) Water Rights Allocation

- Surface water allocation: Allocation principles in the IWEM Plan are : township internal runoff is calculated as the county average runoff multiplied by township areas; Weixi main canal stored water mainly considers nearby townships other townships far from the canal (Fangzhai township and Shoushansi township) do not use canal water. Townships using canal water are allocated water according to each the township cultivated area as a percentage of the total cultivate area of the county; water from the South to North transfer project is equally allocated to townships. These allocations are summarized in Table 5.11.

Table 5.9 Water Resources Supply and Consumption Targets

Township name	Land areas/ mu	Cultivated land/mu	Current water consumption/ mm	Target ET/mm	Should/ mm	Decreasing water/ 10,000 m³
Guantao township	111,000	35,790	545.6 (+79.6)	566	-20.4 (+59.2)	-151.0 (+438.3)
Chaipu township	96,225	80,190	643.2	586	57.2	367.3
Fangzhai township	64,950	43,125	616.1	531.6	84.5	366.1
Weisengzhai township	75,000	58,695	645.8	591.5	54.3	271.5
Wangqiao township	67,500	41,940	627.8	588.1	39.7	178.8
Shaoshansi township	76,500	56,865	646.5	531.6	114.9	586.2
Luqiao township	116,775	69,585	578.7	550.4	28.3	220.3
Xvcun township	76,500	44,250	592.3	584.5	7.8	39.8
Total	**684,450**	**430,440**	**608.8**	**566.2**	**42.6**	**1,942.6**

Table 5.10 Allowable Groundwater Exploitation Under Different Hydrological Conditions

Type of Hydrological year	Water Supply/ mm	Target ET/mm	Surplus / shortage of water/mm	Surplus / shortage of water/ 10,000 m³	Groundwater recharge/ 10,000 m³	Permitted exploitation/ 10,000 m³
Wet year	650.6	584.6	66	3,009.6	7,563.1	4,553.5
Normal year	549.4	549.4	0	0	6,023.5	6,023.5
Dry year	464.4	530.4	-66	-3,009.6	4,759.1	7,768.7
Average year	566.2	566.2	—	—	—	—

- Groundwater allocation: Groundwater is allocated mainly to regions far from the canal or river and where there is shortage of surface water. Township groundwater is allocated according to township area as a percentage of total

county area, as shown in Table 5.11.
- ET water right allocation: ET water right is allocated mainly according to township water supply water from all sources; ET rights are noted in Table 5.11. ET targets change, however, when crop adjustment is taken into account. Crop adjustment is one factor in reducing ET. ET water right allocation after crop adjustment are also shown in Table 5.11.

Table 5.11 Allocation of Surface Water, Groundwater, ET, and ET after Crop Adjustment by Township

Name	Surface Water Allocation/ 10,000 m³	Groundwater Allocation/ 10,000 m³	Target ET/ mm	Target ET after crop adjustment
Guantao township	190.6	923.1	566	495.2
Chaipu township	283.7	800.2	586	588.4
Fangzhai township	94.1	540.1	531.6	534.6
Weisengzhai township	228.1	623.7	591.5	573.3
Wangqiao township	188.1	561.4	588.1	624.8
Shoushansi township	98.4	636.2	531.6	572.7
Luqiao township	267.8	971.1	550.4	605.6
Xvcun township	196.6	636.2	584.5	534.8
TOTAL	1,547.3	5,692.1	566.2	566.2

(5) Water Use Efficiency Plan

To realize groundwater dynamic balance, water savings are an essential component of the Plan. The essential feature of efficient water use is the decrease in ET which, in turn, means real savings of water. The near-term target is to , decrease ET by 42.6 mm. The recommendations in the Plan are based on simulation of a number of possible scenarios involving crop areas, irrigation frequencies, field management, etc.. The Plan identifies the following implementation steps to achieve the ET reduction is target:

- Crop structure adjustment: Decrease wheat and corn by 42,000 mu (15% decrease compared with current planted area); increase cotton area by 42,000 mu. This adjustment in crops reduces crop water requirements, decreases crop ET, and saves 3.804 million m³ of water. These adjustments are distributed to each township in the Plan.
- Irrigation system adjustment: Wheat, corn, and cotton irrigation quotas are decreased by specific amounts (respectively, decrease from 140 m³/mu, 80 m³/

mu, 120 m³/mu to new values of 100 m³/mu, 60 m³/mu, and 100 m³/mu). This further reduces ET saving 14.751 million m³ of water. In some cases this may mean deficit irrigation. However, farmers have been part of the decision to move in this direction in order to save their agricultural futures from a more several water shortage situation in the future.
- Other measures: Farmers are encouraged to adopt other measures for saving water. This may involve water storage, mulching, change of irrigation technique, etc.. Such water saving measures can decrease ET and save a further 1.555 million m³ of water.
- Summary: By implementing the above water use plan, water savings in average years in agriculture are achieved by reduced ET, saving a total of 20.11 million m³ of water and can meet the overall target of 19.425 m³.

(6) Water pollution control and treatment plan

Guantao county is mainly an agricultural county and it is recognized that agricultural nonpoint source pollution may be an important contributor to total water pollution in the county. From model calculations, a decrease in fertilizer use of 40% will greatly reduce nonpoint source pollution by agro-chemicals, however there is serious impact on crop production. Therefore, the Plan uses a decrease of 20% in fertilizer use as the optimal scheme. Additionally, several factories in the county town contributed to point source pollution. In the Plan, nonpoint source pollution control is supplemented by additional point source pollution control.

(7) Integrated management measures

Integrated management measures that are proposed in the Plan include: establish a water rights system; new or revised policies for water resources and water environment management; enforce existing laws; reinforce social measures that improve water management such as community driven development(CDD); a water rights quota management system (demonstrated in this county); improving water charging and measuring facilities; and encouragement to farmers to participate in Water Users Associations (WUA).

5.4.3.4 Conclusion and Summary

Guantao county mainly exploits groundwater with minor use of surface water, and is a typical water-short agricultural county. Because pollution discharge to the boundary river (Zhangwei canal) is prohibited, domestic and industrially polluted water is stored and cleaned locally. Nevertheless, water scarcity seriously affects sustainable socio-economic development. Building an integrated water resources and water environment management system is an urgent task.

The project has used many new concepts to build an IWEM framework, including top-down and bottom-up approaches, extensive involvement of stakeholders (including farmers and WUAs), a resource data system that is shared amongst relevant agencies, and a management framework in which the County Head is actively involved. Directors from the water bureau, environment protection, construction, agriculture, planning commission, and finance sectors are members of the Leading Group and assures cooperation across sectors.

From a technical perspective, use of new KM tools, models and simulations, and a focus on ET management, has resulted in realistic and achievable measures to reduce ET and, therefore, to reduce total water use in agriculture to a sustainable level. Simulation also reveals that fertilizer use can be reduced by 20% without affecting crop production and thereby reducing agricultural nonpoint source pollution. The KM tools have also allowed analysis of different crop scenarios relative to water use and loss, and resulted in a crop structuring scheme that maintains farmers' incomes, yet reduces ET and, therefore, water requirements by the target amount of 19.425 million m^3. Other technical measures include conjunctive use surface water, shallow fresh groundwater, and brackish water substitution of deep fresh water.

Integrated management measures also include: establishing appropriate water resources and water environment policies, revision of the legal framework, inter-sectoral coordination measures, encouraging rural CDD developments, building a water rights system, perfectinga water charging systems together with suitable measuring facilities, and strong support and guidance to farmers to participate in WUAs.

In summary, the combination of measures contained in the Guantao County IWEM Plan decreases ET and reduced water requirements, improves the groundwater situation, decreases pollution and, meanwhile, increases farmers' incomes. By 2020, the Plan can achieve all the targets established to achieve a sustainable water resources and water environment future for the county.

Chapter 6
STRATEGIC ACTION PLAN (SAP) AT THE BASIN AND SUB-BASIN LEVEL

6.1 GENERAL INTRODUCTION ABOUT SAP

Strategic Action Plans (SAPs) are required for the Integrated Water Resource and Water Environment Management (IWEM) at the level of Hai River Basin and the level of Zhangweinan Sub-basin in this GEF/World Bank Project. The difference between the two SAPs is that the former one crosses administrative regions, and the latter one is within one administrative area. IWEMP is one of the core components of the GEF Hai River project which is to provide IWEMP methods and examples for environment projects in China and to improve the environmental condition of the Bohai Sea.

IWEMPs in the project focused on 16county level administrative regions(cities, districts) in the basin-wide. This was based on the fact that it was relatively easier to make a breakthrough from the county(city, district) level than at the whole basin level that involves many jurisdictions. Also, for this project, the large size of the Hai River Basin precluded detailed planning all of the more than 200 counties of the basin. In addition, according to the Ninth and 10th "Five-Year" Plan approved by the State Council for Water Pollution Control in Hai River Basin,more emphasis was put on the pollution control of big cities and big industries at the national level. However, for the vast majority of small and medium towns and small industries, they are usually constrained by funding issues, which is worse in rural areas, and they usually have heavy pollutant discharges. Therefore, by focusing IWEMPs at the county level, and making clarifications about the key performance indicators, it is conducive to promoting pollution prevention and control in the entire basin. Also, by focusing on some counties and small towns in Tianjin Municipality that borders the Bohai Sea, it would have direct and immediate effect in reducing the flow of pollutants into the Bohai Sea. However water management and pollution control cannot be achieved solely by relying on the preparation and implementation of IWEMPs in a few, demonstration, administrative regions; it also requires support and guarantee of the basin-level policies, regulations and management systems. This is also why a SAP for the entire basin was a specific objective of the project. The SAP for the Hai River Basin is based on the main conclusions of the eight strategic studies of the project, as well as the practical experiences in the demonstration counties (cities, districts). It was compiled jointly by the IWEMP teams in the demonstration areas and in Tianjin.

Goals,tasks and specific action plans were proposed in the Hai River Basin level SAP to realize IWEMP in the entire basin.It was proposed to improve the existing laws, regulations, policies, management system, mechanisms, and institutions, and therefore to allocate the tasks of the IWEMPs to each of the project counties(cities, districts), in order to improve the basin and marine ecological environment(to increase environmental water in rivers and wetlands;to stop the overexploitation of groundwater; to improve water environment quality; and to reduce pollutant discharge into the Bohai Sea), to achieve sustainable water use, to reduce evaporation and transpiration(ET)and to reduce excessive wastewater discharges. A "top-down" and "bottom-up" coordination mechanism was formed in the basin in order to achieve the common goals. This would overcome the shortage in the past that the national policies may not be implemented at the grassroots level. SAP at the Hai River Basin level is a major breakthrough in China's water resource and water environment management planning, which will definitely play a significant rolein the overall goals set by the GEF Project of Integrated Water Resources and Environment Management in Hai River Basin.

Zhangweinan sub-basin is within the Hai River Basin, flowing through 13 cities and 70 counties(cities) in Shanxi, Henan,Hebei, Shandong Province,and Tianjin Municipality. The Zhangweinan sub-basin in Hai River Basin suffers from water scarcity, is seriously polluted, has many inter-provincial pollution disputes and threats to downstream water supply safety – all of which seriously threatened regional economic development and social stability. It is the tributary in the basin that has the most serious social contradictions in the whole basin. Furthermore, due to the long-standing lack of unified control of water quality and water quantity, water quality deterioration and water shortages have been affecting each other in a negative way, and this has entered into a vicious circle. Therefore, agreedby the Ministry of Water Resources and the Ministry of Environmental Protection, Zhangweinan Sub-basin was singled out as a breakthrough point for IWEM in the basin, and action plans were specifically proposed to control pollution and strengthen regional co-ordination of water and environment management sectors.

This GEF IWEM Project spanned six years and adapted advanced water management concepts from around the world. The SAP at the levels of Hai River Basin and Zhangweinan sub-basin could also provide practical guidance at the national level.

6.2　SAP AT THE LEVEL OF HAI RIVER BASIN

6.2.1　Basin Features

The Hai River Basin drains much of the North China Plain (NCP) and mountainous areas to the west and northwest of the NCP. The terrain in Hai River Basin is high in northwest and low in southeast, which could be roughly divided into three types of the alpine areas, mountainous areas and plain areas. The basin has semi-arid and sub-humid continental climate, with an average temperature of between 1.5℃ and 14 ℃ over years.

From the point of water resources, the natural characteristics of Hai River Basin are as follows :

Hydrology: The special geographical location makes a dispersed aquo-system, short main streams, and prone to floods. Much of hydrology and river channel network on the North China Plain is artificially controlled by thousands of dams, sluices and transfers. There are no natural rivers as all are heavily diked to prevent flooding. In much of the North China Plain the high degree of groundwater pumping has caused the water table to lie under, sometimes far under, the bottom of rivers so that there is little groundwater recharge of rivers during the winter dry season.

Complex river systems: Although there are separate flows into the sea, there is serious siltation at the estuaries, and it is difficult for floods to pass into the sea. The facts that highlands and depressions of the plains are connected with each other and that the South Canal runs through the north and south makes the rivers in the south flow into the sea concentrated near Tianjin, which makes a lot of pressure on flood control.

There is little rainfall, big changes with dry and wet seasons, intensive storms,

and large spatial and temporal differences across the basin. The average annual precipitation of only 535 mm, with 80% concentrated in the period from June to September. There is also a big inter-annual variability of precipitation, with less that 200 mm of precipitation in arid years, and 1,000 mm or more in wet years. At the same time, it has the highest record in the Chinese mainland both in terms of 24-hour storm intensity and the 7-day storm intensity.

It has water shortages, extreme wet years and dry years, frequent floods and droughts, as well as frequent consecutive dry years. The annual average volume of water resources in the basin is 37.2 billion cubic meters, with a per capita consumption of 272 cubic meters. It is one of the areas with the most severe water shortage. Water quantity varies significantly during the year and among different years. Apart from the serious flooding disasters, there is frequent drought – commonly referred to as "nine years of drought out of ten". Drought also has a particular impact on nonpoint source pollution insofar as lack of runoff reduces mobilization of excessive fertilizer used by farmers.

6.2.2 Socio-economic Characteristics

Special geographic location: Beijing, the national political and cultural center, and Tianjin, the economic, financial and commercial center of Bohai Rim areas are both located in the basin.

Economically developed and densely populated: This basin is one of the most developed areas in the country, with an area of only 3.3% of the country, but a resident population accounting for 10% that of the entire country, and the gross domestic product (GDP) accounts for 14 % of the national level.

6.2.3 Main Problems

6.2.3.1 Technical Problems

Less than 1.3% of the nation's water resources are found in the Hai River Basin, yet the basin is responsible for water supplies for 10% of the national population, 11% of the arable land and 13% of the gross GDP, the water resource development and utilization rate in the basin has reached 98%. With the excessive consumption of water resources, there are a series of ecological problems in the basin, prominently in the following

Chapter 6 STRATEGIC ACTION PLAN (SAP) AT THE BASIN AND SUB-BASIN LEVEL

aspects:
- Dry River courses: An survey about the 21 rivers in the plain areas with a total length of 3,664 km showed that, in the 1950s, the rivers were perennially not dry; in the 1960s 20% of them were dry; 40% of them were dry in the 1970s; up to 60% of the rivers were dry in 2000, with a length of 2,189 km dry, and 12 rivers of them had dry riverbed for more than 300 days.
- Shrinkage of wetlands: The surface acreage of the 12 major plain wetlands including Baiyangdian was 2, 694 km^2 in the 1950s, but only 866 km^2 in the 1970s, which further dropped to 538 km^2 in the year of 2000. There had been a reduction by 80% during those 50 years, and a large number of aquatic organisms had died out.
- Groundwater overexploitation: Before 1965 there was limited extraction of shallow groundwater in the plain areas in Hai River Basin, and land salinization was a major problem at that time. Ever since the 1970s when large-scale exploitation of groundwater was started, groundwater funnels become common. The groundwater overexploitation amounted to 7.92 billion cubic meters in 2004, with 4.16 billion cubic meters of shallow groundwater overexploitation and deep artesian water mining by 3.76 billion cubic meters, which formed 60,000 km^2 of shallow groundwater over-exploitation area and 56,000 km^2 of deep water overexploitation areas, as well as areas with serious ground subsidence. Outflow from the nine main springs had been decreased by 53% over the 50 years.
- Water Pollution Deterioration. There had been water pollution incidents in Guanting Reservoir, Ji Canal and Baiyangdian River in the early 1970s. 28% of all the rivers in the basin were polluted in the late 1970s, 66% of them polluted in 1990, and 72% in 2000. Up to 76% of the areas now have shallow groundwater with quality is worse than Class III, of which 62, 000 km^2 were polluted by human activities.
- Deterioration of the estuarine ecosystems: In the 1950s, annual average inflow into the sea was of 24.1 billion cubic meters, 11.6 billion cubic meters in the 1970s, 6.85 billion cubic meters in the 1990s, and only 3.37 billion cubic meters in 2004. There has been estuarine siltation, serious deterioration of the living environment of aquatic organisms, estuary fisheries relocation and decline in catches.
- Deterioration of ecological environment in Bohai Sea: Bohai Bay is the largest spawning grounds for marine life in the three gulfs of Bohai. With less inflow, the

salinity in Bohai Bay has increased from 25% to 31%. Coupled with the impact of pollution, the spawning and breeding environment in Bohai Bay has been damaged with serious negative impact on fishery production. Ecosystem structure and functions have been destroyed to some extent.

6.2.3.2 Management Problems

In addition to the resource shortage and environmental degradation, there are some serious management problems with IWEM in Hai River Basin, mainly in the following aspects:

- Ineffective systems: For a long time, integrated water resource development and water environment management has been under the charge of several government departments separately, including the water administrative departments at all levels, urban construction departments, and environmental protection departments. This type of decentralized management system resulted in the separation of water use, water management, pollution prevention and flood control, which seriously restricts efficient water utilization and effective protection. At the same time, the basin management agencies do not have the necessary legal status to carry out their functions and therefore cannot effectively resolve the problems with the existing departmental function segmentation and administrative divisions.

- The administrative mechanism is not effective:China has long had a planned economic system; water resource management is based on administrative division. This management approach played an important role in strengthening national macroeconomic regulation and concentrating power. But as the socialist market economic system has been gradually established, the needs of water resource management and water environment management cannot be satisfied by simply relying on administrative means.

- Regulations are not sufficiently effective:There are still many areas for improvement of China's water resource management regulations, for example, there are conflictsand contradictions between laws and regulations in different periodsor by different departments; some terms are not pragmatic; there are a lot of blank spots in the existing water resource management legislation.

- Management mechanisms are backward: Firstly, there are no market mechanisms to regulate water resource management; the water rights and water market systems are still at the initial exploratory stage, which don't have an effective democratic consultation system and cannot adapt to the fluctuating relationship between water supply and water demand. Secondly, the management instruments are not much technically advanced; there are no information systems or decision support systems established for routine water resource development and protectionin the entire basin.

- The right mindset is not yet in place:The traditional concept of "Water Supply determined by water demand" (supply management) has not yet been fundamentally reversed; ecological water use is still neglected; and the idea of "Dry Water Saving" is biased and needs to be urgently transformed.

6.2.4 Tasks and Objectives for the SAP at Hai River Basin Level

In order to solve the problems with water resources and water environment mentioned above in Hai River Basin, to improve the integrated management of water resources and water environment, and to have sustainable economic and social development in the basin, there is an urgent need for a Strategic Action Plan that implements IWEM at the basin level.

6.2.4.1 Tasks for SAP at Hai River Basin Level

The Hai River Basin level SAP, as an important component of this GEF Hai River Project of Integrated Water Resource and Water Environment Management Plan(IWEMP), is designed to develop a strategic Action program(SAP) at Hai River Basin level that focuses on capacity building and integrated management. A subset of the basin-level SAP is the SAP for the Zhangweinan sub-basin that focuses on water pollution control and includes the government's long-term investment plan in Zhangweinan sub-basin. The SAPs focus on capacity building and integrated management. Each of the SAPs should be clear about the specific plansto reduce water consumption and water pollution, to improve the collaboration between different departments, and to establish and improve the local water management.

The Hai River Basin level SAP is mainly about a summary and synthesis of the IWEMPs in the project provinces (municipalities), key project counties (cities, districts), and the sub-basin, the SAP for Zhangweinan sub-basin, and the main conclusions and results from the Strategic Studies and demonstration projects. The SAP also identifies solutions to problems with the institutions, policies, management, monitoring and governance in the basin.

6.2.4.2 Objectives for the Hai River Basin Level SAP

(1) Overall Target

To promote the IWEM in Hai River Basin, to achieve a rational allocation of water resources, to improve the efficiency and effectiveness of water use, to restore the habitat, to effectively alleviate the shortage of water resources, to reduce pollution of Bohai Sea by land-based sources, to improve the water environment quality in Hai River Basin and the Bohai Sea, and to promote sustainable socio-economic developmentin the basin.

(2) Targets for different stages

By taking the year of 2004 as the baseline year, and the year of 2010, 2020and2030 as target years, the medium and long-term level year respectively, targets are set for ET sum control, control of surface water by zone, groundwater overexploitation control, water pollution control, aquatic ecological restoration, wastewater reclamation and reuse, and integrated management capacity building for these different stages.

6.2.4.3 The General Idea for the SAP at Hai River Basin Level

The general idea about Hai River Basin level SAP can be summarized as"one key concept, a core, four systems, and seventasks" , which means as follows:

(1) One key concept

The concept of ET management,water consumption management and total volume control are typically a scientific mindset of "development set by water resources", which should always be at the heart of IWEM in Hai River Basin.

(2) One core

IWEM is the core concept of SAP in Hai River Basin. Future arrangements for the various action plans and measures should be developed based on this core concept.

(3) Establishment of Four systems

These are the legal policy and institutional system, the regulatory capacity supporting system, the construction system, and the safeguard system.

(4) Implementation of the 7 tasks

The seven tasks to achieve sustainable water use are (i) ETmanagement; (ii) distribution and control of surface water; (iii) groundwater exploitation control; (iv) pollution reduction and the compliance of functional zones; (v) ecological restoration; (vi) management capacity building; (vii) drinking water safety.

6.2.5 Principal Recommendations of the Hai River Basin SAP

Based on the eight strategic studies and the water resource plans for the Hai River Basin,and according to the scenarios set by the dualistic core model,indicators were set for actions to accomplish the seven tasks noted above. Strategic action were recommended in seven aspects including water-saving and ET control, the pollutant sum control, the aquatic ecosystem protection and restoration, groundwater protection and restoration, integrated management capacity building and wastewaterr ecycling, etc.. Specific actions are:

Action 1: To enhance the capacity building so as to realize integrated management: this includes policies, regulations, management system and mechanism, monitoring capacity building, management system development, talentteam building, etc..

Action 2: To have surface water control by zones, and to realize regional water balance: The division of the control zones should be in accordance with the traditional water resource zoning and consistent with administrative divisions as far as possible. Based on the inflow into the sea and groundwater control index in the target years, as well as the regional ET targets set in Strategic Study #4 and the conclusions of Strategic Study #7, the control indicators were calculated for different zones in the target years using the Dualistic Core Model and a long series of precipitation data. Engineering and management measures shall be developed by zones.

Action 3: To implement ET control in order to realize real water savings: Analysis is made about the current ET distribution in the basin based on the remote sensing data; ET targets are set and decomposed with the premise of efficient water use in all the areas in the basin without threatening the food production, farmers' income and the environment; the agricultural practices, biological measures, engineering measures and management measures are formulated to achieve target ET; feasibility of the action plans is analyzed in terms of the technical and economic feasibility of the action plan, the acceptance and affordability by users, social and environmental impact, technical support conditions and management tools, and appropriate safeguard measures are developed.

Action 4: To take more efforts for pollution control and to have compliant function zones. Water function zones and the water environmental function zones are integrated and the water quality objectives are established; targets for pollution control are based on the assimilative capacity of COD and NH_3-N by Hai River Basin in the years of 2010, 2020 and 2030, the total emission control is implemented to have compliant water function zones; pollutant reduction measures and safeguard measures are developed in terms of industrial structural adjustment, water saving, pollution remediation, strengthened water reuse, strict emission standards, enhanced monitoring capacity, and so on.

Action 5: To restore the aquatic ecosystem and improve the aquatic ecological environment. The ecological functions of the different rivers are planned and targets set for aquatic ecological restoration. In order to protect the municipal water sources and to repair the river wetland and groundwater, to improve water functions, and to strengthen the allocation of water resources and water function area management, an ecological protection barrier is established up in the upper reaches of the basin to protect city drinking water sources. Aquatic ecological restoration systems are established in the middle reaches of the basin. In the lower reaches of the basin, ecological restoration areas are identified for coastal wetlands and estuaries. Engineering measures are taken to accelerate the restoration of aquatic ecosystem in terms of the river(lakes) environment remediation, channel recovery, wetland recovery, groundwater recovery, ecological environment management in urban rivers and lakes, ecological water configuration, and soil and water conservation measures.

Action 6: To control the exploitation of groundwater and realize the balance between groundwater exploitation and recharge. Standards are set for sustainable development and utilization of groundwater in accordance with the principle of balance between withdrawal and recharge; objectives and programs are formulated for groundwater withdrawal in different target years; measures are recommended to reduce water demand and increase water supply.

Action 7: To reuse reclaimed water, increase water sources and reduce pollution. A unified evaluation system is set up for the classification of the regional wastewater recycling modes by taking into account the need of wastewater reuse, and its feasibility. By summarizing the experiences with planning for recycled water reuse in the eight typical cities implemented in Strategic Study #6, reclaimed water reuse plans are made for other cities in the basin. Based on the perspective of the basin and the ET-based water rights concept, wastewater recycling is analyzed and integrated into the water resource configuration system so that the water use can be optimized and downstream water use is not affected. According to the conclusions of Strategic Study #6 it is recommend to deal with small town sewage by using eco-technologies such as the eco-pond system, underground leaching and filtering treatment system, and artificial wetland systems, and sewage treatment is to be combined with wastewater reuse. The experiences of wastewater management in the coastal areas in Tianjin are summarized and extended to other areas of the basin.

6.2.6 Safeguard Measures for SAP at Hai River Basin Level

Measures are proposed for smooth implementation of Hai River Basin level SAP in terms of organizational security, financial security, technical security, public participation, dissemination and popularization plan, and so on. A list of priority projects is recommended that are supportive of IWEM in Hai River Basin and consistent with the planning in the basin. These projects cover the fields of water conservation, pollution control, ecological protection and restoration, groundwater protection and exploitation control, integrated management and so on.

The State Council issued a document about implementation of the most stringent water resource management system(State Document[2012] No.3). This document specifies that there should be strict water sum use control, to improve the efficiency of water use for the development of a water-saving society, to limit the pollutant assimilation in water function zones and strictly control the total amount of sewage discharge into the rivers and lakes. It also specifies the policies in terms of clarifying water resource management responsibilities and the assessment system, improving the water resources monitoring system, the water resource management system, the investment mechanism for water resource management, and the social supervision mechanism, etc..

The Water Resources Division of Ministry of Water Resources and the Pollution Prevention Division of the Ministry of Environmental Protection jointly issued a key (red) document, which commits to follow up the implementation of the Hai River SAP. The recent projects listed in the Hai River Basin SAP have been included in the relevant planning, such as the Integrated Plan of the Hai River Basin, the Comprehensive Plan For Water Pollution Control and 12th "Five-Year"Plan for the Hai River Basin and in relevant local planning.

6.3 SAP FOR ZHANGWEINAN SUB-BASIN

6.3.1 General Introduction of the Zhangweinan Sub-basin

Zhangweinan Canal is one of the main southern rivers in Hai River Basin for flood discharge and inflow into the sea; it is composed of the Zhang River, Wei River, Wei Canal, Nan Canal and Zhangweixin River, flowing through Shanxi, Henan, Hebei, and Shandong Province and Tianjin City, and finally flowing into the Bohai Sea through Hai River and Zhangweixin River (Figure 6.1). It has a length of 932 km. The river systems are situated in the east of Taiyue Mountain, south of the Fuyang River and Ziya River, and north of the Yellow River and Majia River. The basin is located at longitude $112° \sim 118°$, latitude $35°$ to $39°$, with an acreage of 37,700 km^2, accounting for 11.9% of the total area of the Hai River Basin. Within the basin, the southwest part is relatively high and northeast relatively low. The upstream is located in the Taihang Mountain area, mostly more than 1,000 meters above sea level. The upstream mountainous areas account for 67%, and the middle and lower reaches of plain areas account for 32.5%.

Zhangweinan Sub-basin is a typical sub-basin in Hai River Basin; it has serious problems with water resources and water environment, mainly in the following aspects:

6.3.1.1 Water Shortage and Waste of Water Resources

The annual average total groundwater and surface water in Zhangweinan Canal sub-basin is 6.38 billion cubic meters and 5.16 billion cubic meters respectively, with the total water resources of 10.21 billion cubic meters. Zhangweinan sub-basin has the most serious water shortages in Hai River Basin, with water resources per capita of only 240 m^3 (China's water resources per capita is 2,200 cubic meters, and that in Hai River Basin is 305 cubic meters). The water shortage has seriously affected the socio-economic development of the basin. Mainly due to the reduction of runoff into the rivers, sources for agricultural irrigation and urban water use have gradually shifted to groundwater. Over-exploitation of groundwater has resulted in declining groundwater level to form large funnel areas, which caused uneven ground subsidence and other environmental geological problems. At the same time, there is a serious waste of water resources in the basin. The irrigation coefficient in the basin's irrigation district is around 0.5 to 0.55, and in most areas flood irrigation is still used. The poor irrigation techniques and large irrigation quota result in a huge waste of water resources. With irrational industrial distribution, the industrial water recycling rate is less than 20%, while in the areas with the highest water consumption for every 10,000 CHY output the consumption is up to 700 cubic meters, which is far greater than that in developed countries.

6.3.1.2 Serious environmental pollution, and endangered drinking water sources.

There are in total 31 pollutant outlets in Zhangweinan sub-basin, and the total pollutant discharge into the sea is 810 million tons / year, with 231,000 tons / year of COD emissions, and ammonia emissions of 24,000 tons / year. Ten percent of the total wastewater emissions into the river are by Shanxi Province; Hebei Province accounts for 5%, Henan Province for 70.9%, and Shandong Province for 14.1%. Henan Province is, therefore, a key province for sewage discharge control, as it has insufficient urban sewage treatment capacity and where facilities construction is lagging behind requirements. At present, most of the cities' sewage treatment plants and sewage pipeline construction are very backward. The urban sewage, industrial wastewater, as well as peri-urban livestock wastewater and agricultural nonpoint source pollution result in serious pollution of the rivers.

Water quality in Wei River and its control sections below are of inferior(worse

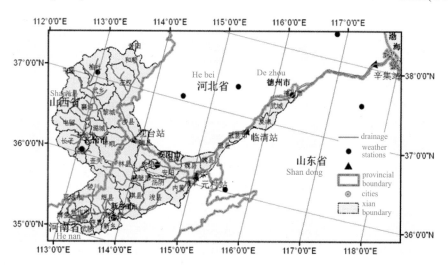

Figure 6.1 The Geographical Location of Zhangweinan Canal Sub-basin

than) Class V; Wei River's headwater of the Dasha River and the main tributaries of Communism Channel, Tang River, Junnei River, and Anyang River are seriously polluted; quality of all the water into the Bohai Sea is worse than Class V, which creates a serious threat to Bohai Sea. River pollution also threatens the safety of the main drinking water sources in the basin, especially the Yuecheng Reservoir and the water source of Jian River in Dezhou city of Shandong Province.

6.3.1.3 Deterioration of Aquatic Ecosystems

Since the mid-1960s, there has been drying up of rivers in non-flood seasons in Zhangweinan Canal basin. In recent years, due to reduced upstream runoff, the middle and lower reaches of the watercourses have been drying up. There has been exacerbated water pollution, environmental degradation, shrinkage of wetlands, and loss of environmental functions. Currently, over 800 kilometers of plain rivers under the charge of Zhangweinan Canal Authority (in turn, under the Hai River Commission) have all become seasonal rivers, and therefore it is impossible to guarantee the ecological water demand of 'normal rivers'.

6.3.1.4 Water Quality Pollution in Offshore Areas

There has been a dramatic reduction of inflow into the sea and large amount of pollutants are accumulated in the inlets. Given the extreme water scarcity in Zhangweinan Canal and over-exploitation of water resources, there is no inflow into the sea in normal year. Zhangweixin River is the main river into the sea in Zhangweinan Canal systems, and there is only inflow into the sea when there is a large flood in the upper reaches. Sea erosion no caused serious damage to the inlets and water ecological environment in Bohai Gulf due to lack of sediment inflow. Moreover, there is only runoff into the sea in a few years, while a lot of pollutants were discharged to the Bohai Bay, which exacerbated the ecological environment destruction in Bohai Bay and resulted in huge economic losses.

6.3.1.5 Serious Pollution in Inter-boundary Sections with Many Inter-provincial Water Dispute.

Inter-provincial water pollution is very prominent in Zhangweinan sub-basin, including the sewage discharged by Henan Province, which is transferred by the Wei canal and Zhangweixin River and causes pollution in Binzhou City in Shandong Province. Wuqiao County, Hebei Province, is impacted by sewage discharge from Dezhou City, Shandong Province. The disputes have not been effectively resolved.

6.3.1.6 Problems with the Management Mechanisms

Currently there remain deficiencies in communication, coordination and information sharing mechanisms between the water conservancy and environmental protection sectors. There is no integrated management of water quality and water quantity. The watershed managementa gencies and local environmental protection agencies have insufficient collaboration. There is a lack of investment for environmental protection management, and the implementation of pollution prevention plans cannot be completed. From the point of IWEM, there remain many deficiencies with the regulations, policies, institutions, systems, and standards.

6.3.2 Zhangweinan Sub-basin SAP

The context, framework and methodology of the Zhangweinan SAP is similar to that of the Haihe River Basin SAP. 2004 was taken as the baseline year, and the years of 2015 and 2020 were taken as the short and long-term target years. Based on the concept of IWEM, internationally advanced technologies for ET management and total volume control are adopted, and a comprehensive, practical and feasible action plan for IWEM was made by use of ET management, remote sensing monitoring, KM system and other advanced methods, in order to achieve the integrated management of water resources and water environment in Zhangweinan sub-basin.

Based on the principle of rehabilitation ofrivers and lakes, the SAP identifies activities that will lead to water resources and water environment security in the basin, to build aresource-saving and environment-friendly society, and to implement energy saving and pollution control. The SAP should focus on solving the pollution problemsof inter-provincial water bodies, as well on protecting the water quality and safety of drinking water sources.

6.3.2.1 Guiding Principles

The SAP process adopted the following guiding principles:
- **Sustainability:** To be based on a realistic and objective analysis of the current situation.
- **Adoption of new technologies:** To focus on the application of new technologies such as ET management, KM system and advanced models to achieve the integration of water quality and quantity management.

- **Horizontal and vertical coordination:** With China's current conditions, the top-down and bottom-up interaction need to be strengthened, as well as horizontal integration and cooperation within and between sectors. Effective coordination mechanisms are required at central, basin, provincial, city and county levels. The needs of local governments and of the public shall be taken into consideration to improve the feasibility and "ownership" of the Strategic Action Plan.
- **Integration with national, basin and local planning:** Based on the outputs of the GEF Hai River Project, focus should be put on linkage with basin and regional plans for water resource management and water pollution control, and linkage with the basin's plan of "energy saving and pollutant discharge control", which shall provide a scientific basis for the IWEM in Zhangweinan sub-basin and its 12th "Five-Year" plan.
- **To realize integrated management in the basin with focus on the pollution control in the basin:** By focusing on the pollution control in Zhangweinan sub-basin, and fully integrating water resource management and ecological restoration, the integrated watershed management can be realized.

6.3.2.2 "One Mechanism, Six Targets"

The SAP for IWEM in Zhangweinan sub-basin lays a foundation for achieving IWEM in the basin. This is to be realized through "one mechanism" to protect the "six targets". "One mechanism" refers to the coordination mechanism for decision-making, between the Ministry of Environmental Protection and the Ministry of Water Resources, the coordination and mutual support amongst the four administrative provinces and one city (Henan, Hebei, Shandong Province, Shanxi and Tianjin Municipality), and coordination between the watershed management organizations and other stakeholders.

The "Six targets" refer to:
- Pollution control and drinking water safety: To reduce the pollutantemissions to the level within the limit established by total load control targets to achieve water quality objectives; to have the water quality in inter-provincial water bodies compliant with the standards and reduce trans-boundary pollution disputes; to implement pollution control in water source areas and have safe watersupply.
- Efficient water use and groundwater restoration: To optimize the allocation of surface water resources in order to meet the requirements by production activities, as

well as domestic and ecological water use; to meet the ET water conservation goals, to reduce groundwater extraction and realize the balance between groundwater withdrawal and recharge, and thereby to gradually recover the groundwater level.

- Ecological security: To ensure the ecological flows, have river ecological restoration and realize healthy river ecosystems.
- Water saving: To realize the agricultural, industrial and domestic water-saving targets, and to gradually reduce invalid water consumption in the basin.
- Target for water quality in the offshore waters: To meet the requirements of water quantity and water quality into the Bohai Sea from Zhangweixin River, and to meet the target of offshore ecological conservation.
- Safe water quality of the South-North Water Transfer: To renovate the aqueducts for water transfer; combined with the pollution control plan of South-North Water Transfer Project, to have the water quality in main delivery canals meet the surface water III water quality requirement.

In addition to these six main targets, the SAP established a number of other objectives. These include the following:

- To promote the all-round implementation of the new concepts of "water saving". This includes the use of new ET control technologies and ET management mechanisms, dissemination and promotion of agricultural water-saving technologies, the implementation of efficient water use measures, and increased water re-use and recycling strategies.
- To enhance the IWEM capacity within institutions of the basin. This included establishing a Leading Group with participation of the decision-making leaders in Zhangweinan sub-basin, as well as the corresponding expert panel, and a coordination management mechanism with joint decision-making by representatives from the Ministry of Environmental Protection, Ministry of Water Resources, and the four provinces and one municipality. Furthermore, the project had joint implementation by project offices in the Ministry of Environmental Protection and the Ministry of Water Resources. An inter-departmental and inter-administrative-region coordination mechanism was established for water resource and water environment management, with water conservation and water pollution control integrated. This coordination mechanisms are to be maintained once the project

is completed and will serve both substantive issues within the basin and as a demonstration to other basins in China on how IWEM can be achieved.

- To significantly enhance pollution control: Industrial restructuring and technological innovation are to be carried out; water pollution control measures are to be implemented in the basin; the point source and non-point source pollution in the basin will be reduced; there will be more sewage treatment facilities in the basin, with better sewage treatment capacity and additional sewage treatment potential; the amount of sewage discharged into the river will be cut to the level within the assimilative capacity of the rivers in the basin, so that water quality at monitored sections meet the required standards. With the national environmental assessment responsibilities, the trans-boundary water quality standards and targets will be met in terms of the sum control of pollutant discharged into the sea, and thereby there will be more compliant function zones. Additionally, nonpoint source pollution will be abated through dissemination of improved fertilizer use technologies, improved irrigation technologies, etc., that will decrease agrochemical runoff into watercourses.
- To have a full implementation of SAP in the basin based on the demonstration county experiences: Through the development and implementation of IWEMPs in Dezhou City of Shandong Province, Xinxiang County of Henan Province, and Lucheng city of Shanxi Province, the demonstration projects are carried out and included inthepriority actions for SAP in Zhangweinan sub-basin.

6.3.2.3 Main Contents of the Zhangweinan Strategic Action Plan

Based on the above principles, objectives and targets, the SAP contains the following principal actions:

- Comprehensive management capacity: The recent and long-term strategic action plans are developed in terms of institutional reform,mechanism establishment, development of laws and regulations, management capacity building, staff training and public awareness, etc., in order to establish an IWEM system in the basin that has advanced technologies, powerful regulations, inter-sector cooperation, information sharing, and scientific decision-making, in order to meet the need of sustainable development of the basin.
- Pollution Control: Near-term and long-term goals are set for water quality of

function zones in the basin, water quality in trans-boundary sections, total load control of pollutant discharge by zones, sewage treatment capacity and processing efficiency, and non-point source control, etc.. Specific action plans are made to achieve these goals.
- Water Saving: Based on water balance, recent and long-term goals are set for water resource utilization, groundwater recovery, ET management and allocation of water resources, etc. in the basin and action plans are developed.
- Aquatic ecosystem security: The ecological water requirement and ecological restoration goals are established in Hai River Basin for targeted sections, and strategic action plans are made to achieve these goals.
- Safe South-North Water Transfer: Based on the South-North Water Transfer Pollution Control Planning, water quality objectives are established for the segment of Zhangweinan canal and strategic action plans are developed to ensure water quality safety of the water delivery line.
- Drinking water safety: According to the recent and long-term objectives for important domestic water sources, urban drinking water sources, and rural drinking water sources, strategic action plans are developed to ensure the safety of drinking water in the basin.
- Water Quality in Offshore Areas: Targets are set for inflow into the Bohai Sea and water quality, and strategic action plans are developed to maintain the safety of coastal waters.
- Industrial restructuring: Short and long-term goals, as well as strategic action plans are developed based on the industrial structure in the basin and that will remediate problems of industrial use of water resources in the basin.

6.3.2.4 Suggestions for SAP at Zhangweinan Sub-basin Level

The Zhangweinan SAP contains specific actions that will meet the goals and objectives of water and environment management within the basin. There are, however, external factors that the SAP cannot control. Below, some of the external factors and recommendation are noted.

- **Integrate the SAP into the basin and localgovernments' the 12th "Five-Year" Plan, and to ensure it is implemented:** The SAP should be effectively integrated with the local government or watershed plans, to incorporate the recommendations into the local planning. The implementation of SAP will help to effectively improve the issues with water resources and water environment in Zhangweinan sub-basin, and thereby to achieve the pollution control objectives and the goal of drinking water safety set by the Ministry of Environmental Protection and Ministry of Water Resources.

- **To effectively monitor and evaluatethe Strategic Action Plan:** Monitoring of the SAP covers M&E for the implementation progress and evaluation and for the achievements of targets. Monitoring and evaluation agencies shall be established to carry out annual evaluation and management of monitoring results, and disseminate evaluation resultsand evaluation reports, as well as to provide guidance for strategic actions.

- **Effective safeguard measures:** Besides integration with the local government planning, the implementation of the Strategic Action Plan also needs safeguard measures in all the following aspects:

  ◎ To have unified leadership with clear responsibilities and accountabilities.

  ◎ Systematic review by local governments and watershed organizations as part of their 12th and 13th "Five-Year" Plans. It is the local governments' responsibility for the implementation of the plans.The provincial and municipal governments need to allocate goals and the tasks to lower levels of city (county) governments, make annual implementation plans, and integrate them into the local annual plans for economic and social development.

  ◎ An accountability system needs to be established, and for those significant environmental accidents due to poor decisions, the responsible officials will be held accountable, as well as those officials who seriously interfere with the normal environmental law enforcement.

  ◎ The watershed management agencies,local governments, and water conservancy and environmental protection departments should have close cooperation, in order to enhance the regulatory capacity and effectively strengthen law enforcement. There need to be more stringent standard systems and improved laws and regulations.

  ◎ To broaden the financing channels and increase financial inputs: With the principle of government guidance, market-oriented and public participation, investment mechanisms shall be established for multi-level investment input by the government, business and social sections, to open up the local, national,social, and international funding sources, including the central investment, provincial and local investment, bank loans, social fund raising and foreign funds, in order to support the project implementation. A national focus is put on Zhangweinan sub-basin, and preferential treatment is provided in funding and project arrangements; local governments at all levels will take them into their own investment plans, and provide preferential policies in terms of the construction sites for pollution treatment facilities, electricity, equipment depreciation, tax, etc.. The key treatment ent erprises need to take active participation in fund raising for its pollution control. Clear pollution control responsibilities need to be taken into account when having corporate restructuring. Professional companies are encouraged to take part in the construction or operation of pollution control facilities. Efforts need to be made to have IWEMP included in local planning and to have the national and local funds to support its implementation. Efforts will be made to raise investment for the Phase II of GEF Hai River Project.

◎ To strengthen scientific and technological support to improve the management capacity: IWEM in the basin requires strong scientific and technological support, especially for the SAP where a variety of model are used for simulations, such as the SWAT model, environmental assimilative capacity calculation model, the pollutant abatement model, and so on. With the use of these models the efficiency of the strategic plans are improved. As examples:

◆ The application of advanced ET technologies makes GEF Hai River project very different from conventional water resources management projects. It is necessary to establish an ET center with the relevant research and development investments. Given the very large amount of water wasted across the basin from non-beneficial ET, this is a key investment.

◆ KM has been demonstrated to be a set of key technologies for analysis and operation of water and environment management systems. Data sharing needed to be further encouraged in the future between the environmental protection and water conservation sections, and at different levels. The GIS-based management information system for integrated water quality and water quantity management in Zhangweinan sub-basin is established. Decision-making support modes are developed for surface water and groundwater quantity and quality management, as well as an complete database and geographic information system (GIS). All these systems need to be maintained and continually improved in practice in order to meet the demand for integrated management.

- To carry out assessment of the plans and promote the implementation of planned actions: The system of annual appraisal of plans will be carried out. Governments at all levels in the basin need to establish statistical, monitoring and evaluation systems that are compatible with total load and volume control. Annual analysis and evaluation should be made for the implementation progress of the plans, water quality, total amount of emissions, ET reduction and environmental management, etc., and necessary adjustments shall be made to the schedule timely, to promote the implementation of the action plans.

- Mechanism for public participation and advocacy: In order to ensure smooth implementation of the SAP, educational programs for public participation and extensive publicity should be developed and implemented to mobilize the enthusiasm of the whole society, to promote the implementation of the planned tasks. By setting up hotlines, the public mail box and carrying out social surveys, etc., public feedback and timely solutions can be attained about environmental issues and will further social harmony.

- To establish the mechanism for public feedback and to ensure the implementation of SAP: Departments of environmental protection, water conservancy, construction, and health care should cooperate closely with the establishment of environmental information sharing and transparent management system. Citizens, business sections or other organization saffected by water pollution can file for compensation through legal means so that their legal environmental interests can be protected.

Chapter 7
SUMMARY OF PROJECT OUTOMES AND EXPERIENCE

7.1 CONTEXT OF THE PROJECT

During the period 2004 and 2010, with financing from the GEF and the Government of China, the China GEF "Hai River Basin Integrated Water Resources and Environment Management" was carried out. The project was implemented jointly by water resources and environment protection sector from Hai River Management Institutions, Zhangweinan Sub-basin, Hebei Province, Beijing Municipality, Tianjin Municipality and 16 project counties (cities, districts), under the joint project management of the Ministry of Water Resources (MWR) and the Ministry of Environment Protection (MEP), with technical guidance from the World Bank. The goal was to promoting integrated water resources and environment management for the Hai River Basin, to reducing the water pollutants load volume into the Bohai Sea from this river basin, to reduce groundwater overdraft, and to improve the basin and Bohai Sea environment.

This GEF Hai River Project is the first project in China with integration of land and sea, water resources and water environment. Because basin management is fragmented amongst different sectors, it is difficult from an institutional perspective to achieve effective integrated planning and management of water resources, pollution control and to achieve a satisfactory marine environment in the long term. Usually, sector targets and plans overlap and/or in conflict. This lack of coordination leads to IWRM failure and is the underlying context of this project. Because of the extreme problems of water scarcity and high levels of water pollution and integrated approach is needed in China – something that has been recognized by Chinese and foreign academics for a very long time. The water resource and water environment situation for the North China Plain which occupies a central position in China's political, economic and social life and which is largely drained by the Hai River, is now so critical that this Project was developed by the Chinese Government with the World Bank to demonstrate an alternative water management model for China.

7.2　MONITORING AND EVALUATION

The Project developed a full M&E program which included independent audit of the M&E results on a regular basis. An M&E protocol was developed for each project implementation office (for each of the 16 demonstration counties, for the Hai River Basin Conservancy Commission, for the Zhangweinan Sub-basin office, and at the central level). High level indicators reflected that key outcomes of the Project (see section 7.4). The M&E indicators included lower level indicators that reflected the particular requirement in each of the demonstration counties (cities, districts). All M&E programs included activity indicators (process indicators) that allowed roll-up of status of implementation as a measure to evaluate overall implementation progress by each World Bank supervision mission. M&E performance was also reviewed by the International Expert Panel that regularly provided guidance to the Project.

7.3　PROJECT OBJECTIVES

Using best available concepts and technologies the PAD for this project identified the following main objectives:

- Introduce and demonstrate advanced basin management concepts that can integrate the relationships amongst water resources utilization, pollution discharge and sea response, and to develop an Action Plan, with costs, for basin water resources utilization and water and sea environment protection.
- Established an integrated water resources and environment management system at the basin level that will lead to water conservation, flow increases, decreases in nitrogen and phosphorus wastewater loads.
- Established inter-agency cooperative mechanisms amongst water resources, environment protection, urban construction, agriculture and oceanic sectors.
- Established a Knowledge Management (KM) system that facilitates joint decision making and which is shared by all relevant departments at all levels of government.

- To introduce and implement the concept of "water-demand" management, with water-consumption control to replace the existing "water utilization control" paradigm; this requires introducing evapotranspiration monitoring and control (ET management) at all levels in the basin in order to make quantitative savings of ET that is lost from the basin.
- Establish targets for reduction in groundwater overdraft and of pollution loads using targets years of 2010, 2015 and 2020.
- Formulate and implement an Action Plan using "top-down" and "bottom-up" activities so that all stakeholders are involved.

7.4 MAIN OUTCOMES OF THE PROJECT

The GEF Project Appraisal Document (PAD) established a set of qualitative and quantitative targets to be completed in the project. Using these plus additional targets established during the inception period, full monitoring and evaluation process (M&E) was established to track project progress on six-month interval basis and to report to the Executing Agency (World Bank), the Chinese Government, and to the GEF. The key performance indicators set out in the PAD and results of the M&E of performance results are shown qualitatively in Table 7.1.Table 7.2 summarizes the main quantitative targets and indicator completion status at the end of the project (2010).

Pollution Reduction: The key performance indicators called for a 10% decrease in pollution loadings by 2010 relative to the baseline year of 2004 in the project counties and at least one coastal counties. In fact, as shown in Table 7.2 all key performance indicates were achieved and, in many cases greatly exceeded. All pollution indicators show a downward trend since the beginning of the project. The project does not take all the credit for this insofar as part of the cofinancing of this Project included planned expenditures on increased sewage infrastructure, especially in the Tianjin area.

For the sediments alongside the riverbeds and for riverbank soil, sediment removal and riverside rehabilitation treats the pollutants. As a result of this situation, COD and NH_3-N discharges into sea from these sources are 9,845.3 tons per year and 856.7 tons per year respectively in average year.

Table 7.1 Seven Key Performance Indicators and Project Results

	INDICATORS	RESULTS
1	Establishment of a functioning inter-agency committee at the county level, resulting in improved cooperation and integration of Water Resource Management (WRM) and pollution control activities with support from upper levels [prefectures, provinces, Hai Basin Commission (HBC), Zhangweinan, Ministry of Water Resources (MWR), Ministry of Environmental Protection (MEP)]	A functioning inter-agency coordinating committee was established in each demonstration country and was the key factor in coordinating the activities of water, environment, construction and agricultural agencies. These coordinating committees were fully supported by higher level organizations as required in the indicator
2	Achieve the adoption of improved WRM and pollution control approaches at the county level by institutions implementing Integrated Water and Environment Management (IWEM), including evapotranspiration (ET) management and knowledge management (KM), water rights and well permit administration, and discharge control, with support from upper levels (prefectures, provinces, HBC, Zhangweinan, MWR and MEP)	Improved WRM and pollution control planning was fully achieved; this included operationalizing an ET management systems as the principal vehicle for controlling the water balance at county and lower levels and for achieving "real" water savings. Well permitting was implemented in the chosen demonstration county. The KM system was fully developed and the platform was provided to each demonstration country for IWEM purposes
3	Achieve the adoption of improved WRM and pollution control approaches at the county level by institutions implementing Integrated Water and Environment Management (IWEM), including evapotranspiration (ET) management and knowledge management (KM), water rights and well permit administration, and discharge control, with support from upper levels (prefectures, provinces, HBC, Zhangweinan, MWR and MEP)	All measures proposed for demonstration counties (cities, districts) were adopted and implemented according to the PAD. IWEM Plans have been adopted and implementation has begun. The coordinating mechanisms noted above have committed to maintain the IWEM process for the long term
4	Reduce discharge pollution load by 10% in pilot counties and coastal counties	The pilot counties (cities, districts) achieved 10% reduction in sewage volume and of main pollutants. Sewage and pollutants discharge have declined year by year after project implementation; by 2009, sewage, COD and NH_3-N targets in all 16 project counties (cities, districts) had been met. As shown in Table 7.2, the PAD targets were substantially exceeded

Chapter 7 SUMMARY OF PROJECT OUTOMES AND EXPERIENCE

	INDICATORS	RESULTS
5	Reduce groundwater overdraft for irrigation purposed by 10% in pilot county (Xinxiang County)	The project counties (cities, districts) have achieved the target of 10% reduction in groundwater over-exploitation. Deep and shallow aquifer groundwater met or exceeded the reduction target by 2009 (Table 7.2)
6	Reduction of pollution loading to Bohai Sea from at least one Tianjin small city by 10,000 tons of COD and 500 tons of NH_4 annually	During the project the Hangu (Yingchen) sewage treatment was established, leading to the reductions that satisfy the project target. Details are presented in Table 7.3. As shown in Table 7.2, the PAD targets were substantially exceeded in the three pilot counties that discharge to the Bohai Sea (Baodi, Ninghe, Hangu)
7	Disposal of 2.2 million cubic meters of contaminated sediment from the Dagu canal in an environmentally safe manner, and achieve a one-time reduction of 10,000 tons of oil, 2,000 tons of zinc, and 5,000 tons of total nitrogen	Calculated values from sediment sampling indicate that 6.26 million cubic meters of contaminated sediments were removed; the chemical decontamination was slightly below target for zinc but the rest met the target. Given the uncertainty in extrapolation from sediment samples it is considered that the targets were achieved. Details are provided in Table 7.4

The pollution data need some explanation. As most riverbeds of Hai River Basin are mainly seasonal channels or are dry, they transport little water for much of the year. Therefore, because of absorption and seepage effect of sediment and riverbank soil alongside rivers, most of the pollutants do not reach the Bohai Sea from counties that are not located along the coast. The exception to this is in years when there are large floods that carry inland pollution loads downstream to the ocean. Such floods are relatively rare. As the riverine environmental quality suffers badly from inland pollution loads, the Project adopted two types of pollution load control targets – the first is for demonstration counties located away from the coast insofar as riverine ecosystem health is a long-term goal of the government, and the second is that which is delivered to the sea from coastal locations (counties, districts, cities). Locations included in the Project for load to sea targets are Baodi District, Ninghe County and Hangu District in Tianjin Municipality. Another type of pollution load to the sea is that from the Dagu sewage canal which is a

Table 7.2 Macro Target Indicators Status at End of Project Based on Monitoring Results

Monitoring period / Monitoring project		ET /mm			Groundwater Over-exploitation/ 10,000 m³		Pollutants discharge into river from pilot counties¹			Pollutants discharge into Bohai Sea		
		RS monitoring	Water balance calculation	precipitation	Shallow Aquifer Ground water	Deep Aquifer Ground water	total sewage/ 10,000 tons	COD/ tons	NH$_3$-N/ tons	Total sewage/ 10,000 tons	COD/ tons	NH$_3$-N/ tons
Actual monitoring Values	Baseline(2004)	**586.9**	**603.6**	**570.2**	**41,885.9**	**27,201.2**	**38,701.4**	**108,372.7**	**12,152.9**	**6,244.1**	**18,661.4**	**1,821.6**
	2005	516.1	549.6	505.7	22,444.0	21,302.6	36,829.8	101,790.3	11,196.6	6,634.8	20,096.0	1,399.6
	2006	515.9	545.6	439.0	43,502.0	21,560.1	29,902.1	83,728.3	9,453.1	3,595.1	8,523.0	913.2
	2007	535.0	558.3	496.9	25,246.1	20,710.5	27,792.2	64,143.5	8,256.2	3,379.4	8,927.6	923.8
	2008	545.8	587.7	587.8	-101.4	18,350.0	27,623.5	59,998.7	7,380.5	3,343.8	7,968.6	855.0
	2009	563.4	534.2	540.4	21,169.6	14,527.9	24,960.9	40,133.1	5,562.9	2,676.7	6,613.4	750.0
	2010	524.0	549.6	518.4	15,370.9	14,733.5	25,767.3	38,614.8	4,665.3	5,531.0	6,951.0	804.7
% Reduction in 2010 relative to baseline		10.7%	8.9%	51.8	63.3%	45.8%	33.4%	64.4%	61.6%	11.4%	62.8%	55.8%
PAD Target value		579.1			37,697.3	24,481.1	36,731.8	103,677.9	11,492.4	5,812.3	17,642.1	1,721.5
(% Reduction in 2010 relative to PAD target)		9.5%	5.1%	—	59.2%	39.8%	29.9%	62.8%	59.4%	4.8%	60.6%	53.3%
	Long-term target (2020)	543.2		—	0.0	0.0	19,005.6	61,424.7	5,548.2	1,926.0	8,468.3	820.5

(green = exceeded target or baseline)

Chapter 7 SUMMARY OF PROJECT OUTOMES AND EXPERIENCE

major repository of contaminants in the canal sediments that continuously leak to the sea (Table 7.4)

Summary: As shown in Table 7.2, project monitoring and evaluation demonstrate that this GEF Hai River Basin project completed all of contents during the implementation period. It also achieved all implementation planning tasks in accordance with Project Agreement and Project Appraisal Document.

Table 7.3 Small Township Sewage Discharge Indicators (load to Bohai Sea)

Monitoring value / Main pollutants	Baseline year level (2004) Treatment amount/(ton/year)	After establishment(2009)		Target value Annual treatment amount/(ton/year)
		Daily treatment amount/(ton/day)	Annual treatment amount/(ton/year)	
COD	0	27	9,855[1]	10,000
NH$_3$-N	0	1.7	620.5	500

1 Within measurement error.

Table 7.4 Dagu Discharge Ccanal Pollutants Reduction (pollutant load to sea)

Monitoring content / Year	Pollution sediments (ten thousand m^3)	Oil (ton)	Zinc (ton)	Total Nitrogen (ton)
2004 (Baseline)	0 reduction	0 reduction	0 reduction	0 reduction
2009	626	28,670	1,820	13,378
Reduction amount	626	28,670	1,820	13,378
Target value	220	10,000	2,000	5,000
Achieved	**Completed**	**Completed**	**Completed**[1]	**Completed**

1 Within measurement error.

7.5 PROJECT IMPLEMENTATION EXPERIENCE

Because this was the first large-scale IWEM project in China and also the first that dealt directly with the traditional institutional separation of water from water environment, the following text describes in more detail some of the important experiences gained from this GEF Project. This includes elements that are useful for conducting integrated water resources and environment management and strategic action planning (IWEMPs and SAPs) in China. In addition, it points out potentially uncertain elements and difficulties during the process of planning. It shares successful experiences on public participation, "bottom-up, top-down" and vertical-horizontal integration working approaches.

7.5.1 Cross-departmental Cooperation Mechanism

7.5.1.1 Background

The single largest barrier to IWEM in China is the lack of coordination and, often, even of significant cooperation, between the Ministry of Water Resources and the Ministry of Environmental Protection that are the two most important institutions that manage water quantity and water quality respectively. This institutional separation reflects, in part, (a) Chinese law in which each ministry has its own laws and that are often not well coordinated in terms of legal text and typically lack a high degree of specificity on mandate, terminology, etc.; (b) the fact that, historically, the Chinese management system is vertically structured in which "power" is devolved downwards, not shared horizontally. This separation of authority and lack of horizontal cooperation extends downwards to subordinate organizations at provincial, county and township levels, as well as to the professional agencies attached to each of the main ministries and to river basin commissions (that belong to the Ministry of Water Resources). In addition to these two ministries, other institutions also have important roles in water management, including agriculture and construction ministries and their provincial departments.

Because stable and effective horizontal and vertical integration at all levels (ministry, provincial, Hai River Basin and Zhangweinan sub-basin commissions, municipal, county, etc.) was a prerequisite for a successful project, the World Bank requested that the Chinese government submit a formal project cooperation mechanism document as one of the legal documents for the project. This document provided the basis for a new and effective set of cooperative arrangements amongst all the major players at all levels of the project (hereafter referred to as "project cooperation mechanism").

7.5.1.2 The Project Cooperation Mechanism

This is comprised of seven parts:
- The goal of coordination is well defined so that the boundaries, obligations and benefits of cooperation are accepted by each of the parties.
- Each relevant party to the cooperation mechanism was fully defined from central to local levels.
- The basic concepts for project cooperation mechanism were defined, including a

Chapter 7 SUMMARY OF PROJECT OUTOMES AND EXPERIENCE

realistic work style, mutual respect, seeking common ground while accepting the existing differences, mutually promotion, timely communication, labor allocation with individual responsibility, and mutual coordination. It requires timely communication and coordination at all levels of the Project. The CPMO (Ministry of Water Resources and Ministry of Environmental and Protection) are responsible for implementing the cooperation concept and are entirely responsible for project management, supervision, and all executive actions.

- The Project Management Office (PMO) at each level accepts responsibility for a uniform approach to project management so that there are similar requirements for all PMOs extending from ministry to county levels. This included the "leading groups" at each level that would be responsible for ensuring appropriate leadership by senior local officials (e.g. bureau directors, mayors, etc.). At the highest level, the Central Project Management Office (CPMO) was established in each of MWR and MEP to ensure overall coordination, coordinate M&E at all levels, control fund expenditures, provide oversight on contracting and sub-contracting, and maintain the management information system (MIS) and special website.

- Responsibilities of each of the parties was clearly defined. This included both substantive responsibilities involving project outputs and outcomes (who is responsible for what) but also responsibilities for reporting, financial management, M&E obligations, procurement, etc.. The proportion of responsibility allocated to each of the ministries also defined the responsibility for co-financing in the same proportion. The responsibilities followed ministerial mandates so that MEP was mainly responsible for the pollution and environmental aspects of the project, whereas MWR was mainly responsible for the water resources side of the project. The cooperation mechanism ensured that these responsibilities did not remain isolated within the respective ministry.

- To deal with specific coordination or mandate problems as these arose, an arbitration mechanism was established. As these problems generally originated at lower levels (usually due to lack of clear understanding of some issues) the project adopted a "bottom-up" approach in which issues are brought to the attention of higher levels of project management and, if required, to the top level arbitration institution operated jointly by MWR and MEP. Issues involving legal covenants of the project certain issues could be submitted to the international department of the Ministry of Finance (MoF). For technical issues, arbitration was bottom up, from the joint expert group at one level to the joint expert group at the next level. The most senior technical arbitration mechanism is the Central Joint Expert Group. Technical issues most frequently reflected not "who should do it", but exactly what should be done. Because many parts of the Project were highly innovative and very complex, the Central Joint Expert Group was heavily involved in defining the way forward on complex technical issues.

- Certain issues that were potentially contentious and involve the legal mandates of MWR or MEP required special attention. These had to be resolved to the mutual satisfaction of both ministries. These issues included:
  ◎ Environment impact evaluation that is required under the environment impact evaluation law.
  ◎ Because Project implementation activities had, in some cases, environmental impact, there was a requirement for joint work that could involve construction, water and soil conservation, water resources and ecological management in the project areas, etc..
  ◎ The calculation of assimilative capacity in waters required decisions on how this should be done.
  ◎ Coordinate how monitoring of water function zones and water environment zones will be done and how to share these data; also, to coordinate water quality targets in project areas step by step so as to meet water quality requirements in water function zone and also meet the target of environment management.
  ◎ Coordinated approach to water intake permission and pollution discharge permission.

7.5.1.3 The Impact of the Cooperation Mechanism to the Project

Project cooperation involved a very wide variety of working levels – MWR, MEP at eh national level, HWCC, ZWN canal sub-basin at the basin level, Beijing and Tianjin municipalities, and the demonstration counties. The project cooperation mechanism has been implemented for 7 years, and has demonstrated to the rest of the country how integrated water resources and environment management can be effectively coordinated.

More particularly, the following points are germane:
- Project cooperation mechanism successfully promoted project progress so as to ensure the significant outcomes. Most importantly, this included:
  - 17 IWEMPs involving one municipality, 3 provincial levels, 13 county levels, including Tianjin, Beijing, Hebei, Shandong and Shanxi;
  - The "Hai Basin Knowledge Management System and Data Sharing Agreement" signed between the Pollution Control Department of MEP and the Water Resources Department of MWR.
  - Promote substantial cooperation between environmental protection departments and water conservancy departments at various levels during the project implementation period; each consultant service contract was approved by both sides and the results review jointly.
  - Joint technical joint review system and work inspection management system; it also continuously has updated project cooperation mechanism due to project development situation so as to play an effective role for project implantation.
- The Project cooperation mechanism established a brand new cooperation platform so as to fully express participants willingness to participate and to exchange information effectively. The outcomes can be shared amongst professional teams involving hundreds of people.
- Because of the high cost of managing a large and complex project, the cooperation mechanism reduced redundancy in project management and therefore reduced management costs.
- Project cooperation mechanism was the first example of integrated cross-ministry management process, from planning approval, bidding procurement, sub-contract, financial management, finance allocation, information management, investigation assessment and to final outcomes review. There is no precedent for this in Chinese experience.
- "Cross-ministry" is an area of current public administrative research.The Project cooperation mechanism provides a good example, promotes good government management performance and reduces cost.
- The Project has demonstrated that constructive cooperation between environmental protection and water conservancy departments is not only possible but can be

developed and maintained as a practical working principle.

7.5.2 Key Innovations

The Project has produced many innovative methodologies and technologies to achieve success. The key innovations are noted below. The technology innovations are discussed in more detail in Chapter 3.

- **Bottom-Up, Top-Down, and Horizontal management mechanisms**

China is well known for top-down ("command and control")management; however, the country recognizes the need to receive inputs from lower levels of society. In this Project, because water resources directly impact people's lives, the Project was developed around both top-down and bottom-up approaches to ensure that the flow of information reflects both mandated issues (top-down) that are conveyed downwards from ministries and provinces, and local aspirations and concerns (bottom-up). Bottom up in this project involved both the local government's interests and the interests of citizens such as farmers in consultations on topics such as water in irrigation and ET management. Additionally, the Project used "Horizontal coordination" – meaning, an integrated management process involving two or more units at the same level of government.

- **Integrated water and environment management**

This has been described above, but is highly innovative for China and demonstrates how the usually contentious relationship between water and environment can be effectively managed. It is perhaps best illustrated in the major metropolis of Tianjin which suffers extreme water scarcity and high degree of water pollution. Since 2004, with this Project, the water resources bureau and environmental bureau have developed an overall, integrated water resources and environment management planning framework that is unique amongst major cities of China. The Tianjin Leading Group for this Project included the municipal financial bureau, construction commission, water conservancy bureau and environmental protection bureau as well as leaders from other relevant departments. A joint Project management office was established under the water conservancy bureau and environmental protection bureau. Close coordination was developed between the Tianjin Project Office and project offices managing the pollution control construction and for the Dagu Canal remediation project, so that these two latter

Chapter 7 SUMMARY OF PROJECT OUTOMES AND EXPERIENCE

activities were fully integrated into Tianjin integrated water resources and environment management planning (IWEMP).

The Tianjin IWEM activities had the following major successes:

- Joint project management including activity development, technical inspection, vetting of contractors and consultants, progress reporting and quality control. This is "first" for China.
- IWEM jointly developed and subsequently approved by the Tianjin Development and Reform Commission in 2007. This step guarantees plan implementation.
- Public participation was central to development of water-saving plans at the field level. By the end of the project training had been provided to many farmers groups (water users associations – WUAs) and more than 100 WUAs have been formed within Tianjin Municipal area. Farmers provided important feedback on many issues that helped form the IWEM plan. WUAs were used as a platform for farmers who are major water users, to participate in planning and in management water and environment resources management.
- Revision of water function zones was carried out jointly by water and environment sectors. This removed a major problem for effective water management by eliminating the dual function zoning system.
- The KM system was developed and adapted for Tianjin purposes and is now used by both ministries.
- A joint expert panel of external experts in water and environment was established under the joint direction of the water conservancy bureau and the environmental protection bureau. Never before has such a joint expert panel been used in China.
- Institutional reform: Integration of water-related construction, highway drainage, and water management was consolidated in 2010 into the new water resources bureau with the agreement of the State Council. This represents consolidation of management for what had been overlapping and often competing jurisdictions.
- Regulatory reform: During the Project, Tianjin improved its system of rules and regulations. Tianjin has revised and implemented "Water law by the People's Republic of China methods" "Tianjin urban water supply water use

regulations""Tianjin water pollution preservation management method" and "Tianjin cleaner production improvement regulations" which, with other regulations, provides for a water resources and environmental management regulation system framework. It has also decreed strict compliance to technical standards – Tianjin issued and implemented "Tianjin sewage integrated discharge standard in local areas" at the beginning of 2008. It also decreed local requirements under the national sewage integrated discharge standard, and revised the discharge criteria for industrial discharges.

◎ Tianjin municipal government designated exclusion areas for groundwater extraction and established verifiable quotas on groundwater extraction.

◎ Remediation of polluted rivers was begun in 2008 under an action plan that focuses on 39 seriously polluted rivers and/or river reaches. At the time of writing, this action plan has been completed for 10 rivers in the central part of Tianjin treatment, with 29 additional rivers under treatment.

◎ Optimization of water use: This encompasses the entire municipality with a particular focus on water use in agriculture. A water charging system is now being explored as a vehicle to promote water saving by farmers, as well as other measures such as crop restructuring, ET control, water-saving irrigation applied to some 65.8% of all irrigated area (by the end of 2008), etc..

(1) Holistic approach to river basin management

The Hai River Basin covers 7 provinces (autonomous region, municipality), 24 cities, 219 counties (cities, districts) and 85 national (hydrometric) cross-sections with 7 control areas, 73 control unit scheme. The water resources shortage for the Hai River Basin and serious water pollution are huge issues that have been difficult if not intractable to tackle under the conventional form of basin management. Through an integrated approach, and using modern and innovative technical tools, the Project demonstrated how solutions can be found for the basin as a whole. More importantly, the Project demonstrated that outcomes take time but can be achieved step-by-step with agreed and technically defensible targets that make sense in the current context of China. Many of the recommendations of the Project have, therefore, been incorporated into the 12th "Five-Year" Plan for this basin.

Water quality and water pollution is a highly sensitive and contentious issue across the basin with many serious disputes between provinces. Unlike previous five year plans in which pollution targets reflect individual provincial targets that are not integrated across the basin (and therefore fail to achieve in-stream targets), the Project has determined an initial 15 priority control units in the whole basin, including one in Beijing, one in Tianjin, 5 in Hebei, 3 in Henan, 3 in Shanxi, 1 in Shandong and 1 in Inner Mongolia autonomous region as a basis for basin-wide pollution control.

(2) Knowledge management (KM) system

While KM systems for river basins are common-place worldwide, few have the capacity to integrates and mobilize such a wide diversity of knowledge, data, and information, and to create outputs that link the different parts of the hydrological cycle, ET monitoring, pollution management options, ecological requirements, etc., together. The fact that this system can work in a basin of unprecedented complexity is huge credit to the KM developers. The dualistic core model for the project is unique; it takes full consideration of natural runoff, hydropower, and human planning/management/ water resources activities and other elements. It can be divided into two components representing natural and human interactions. It analyzes Hai River Basin water resources development utilization, water function areas and water environment function areas, river discharge drainage, Bohai Sea cross-section, and determines main system control cross-sections.

(3) River reach coding system

Like most river basins in China in 2004, each of the ministries of water resources and environmental protection had their own system for coding river segments. To some extent the existence of these two systems represented the inability of the two ministries to cooperate in river management. This was recognized in the Project as a major barrier to achieving an integrated approach to river management. Therefore, a river reach coding system, drawing inspiration from the United States STORET river reach coding system, was developed as a means of integrating the two separate systems into a common coding framework. This was a unique response to a nationally difficult system of coding. This new system can access data that is held by either ministry so that water and pollution information can be brought together for integrated decision-making. The system has 204 water environment function areas and 138 water function areas; with reorganization it will form 729 sections in the network.

(4) ET, water balance, and "real water savings"

The Project adopted the principle of "real water savings" which means that water use efficiency cannot be defined solely from the perspective of a single sector. For example, water loss from a canal is considered to be "lost" by irrigation engineers, but in fact, this "lost" water is recovered in the groundwater and is available for some other use such as well water used for irrigation. Real water savings are savings from water that is lost to the basin, not to a particular sector. For the most part, water that is lost to the basin is in the form of non-beneficial evapotranspiration (ET). Because evaporative loss in this semi-arid basin is very large, any significant saving in water use must initially focus on controlling the loss from ET. Therefore, the Project developed the methodology for measuring and managing ET using ET quotas that are passed down to county and to farm levels. This approach has never been used in this way anywhere in the world. ET management is achieved in three areas: (i) strengthening water management;(ii) changing crop structure and land use, and (iii) use of best management practices to reduce soil evaporation (new types of irrigation, mulching, etc.).These include scientific irrigation, improving seed, changing crop structure, greenhouses, straw covering, lying fallow and no-till. The Hai River Basin ET Management Center and remote-sensing ET technology now continuously applies ET water management to check and ratify permitting that determines allowable groundwater exploitation. ET management focuses on water consumption instead of water demand and is a major take-home lesson from this Project.

(5) Data sharing

The lack of data sharing between the water and environment sectors has been a major problem for effective water and environmental management in China. Therefore, the MWR and MEP formally agreed to share relevant data using a common KM platform. Data are stored and accessed through the river coding system (see above). The basis of the data sharing agreement may be instructive to other basins in China, as follows:

- Within the scope of defined data and information the water resources and the environment protection departments agreed to a free exchange of data through a data exchange mechanism.
- Within Hai River Basin Level KM system under the management of Academy of

Environment Planning under the Ministry of Environment Protection, and the Hai River Conservation Commission of the Ministry of Water Resources, a shared database was configured for each group. These the two technology units exchange the stored data collected by Hai River Basin water resources and environment protection departments at the first day of each month for the use of both ministries.
- Twice a year the supervising leaders from both parties jointly meet to review and summarize the implementation of the agreement and to study and deal with any problem issues.

(6) Pollution control
- Pollution Standards: The paper-making industry is a key factor in the control of industrial pollution in the Hai River Basin. Using Zhangweinan Sub-basin as one example, there are serious problems with non-compliant discharges, an excessive number of small enterprises, and large number of straw-based paper production factories. If current enterprises could meet the requirements of COD discharge concentration under the new paper-making standard the COD discharge volume from paper-making enterprises will dramatically decrease. Therefore, improving the discharge standards for various industrial groups is one important approach for industrial pollution control. However, the Project also found that industrial compliance to discharge standards is low, indicating that new standards together with greater compliance can bring considerable relief from water pollution. The Project also found that because of very low flows across the Hai River Basin, area-specific standards are required as have been implemented in the Lake Tai area and in Beijing and Tianjin Municipalities.
- Industrial Restructuring: The Project established that there is a mis-match between water use and pollution by industries on the one hand, and value-added from various industry groups. This suggests that industrial restructuring where high-value, low water use and low polluting industries should be encouraged, whereas low value, high polluting and high water using industries should be discouraged. As examples, the paper-making, chemical and pharmaceutical industries in Xinxiang City account for only 20.6% of total production value while the COD and NH_3-N amount discharged from these same industries account for 75.9% and 88.6% respectively of total COD and NH_3-N discharges. Especially in the paper-making sector, with only

6% in production value, the COD and NH_3-N generated accounts for 67% and 45% of the total. In Dezhou City, paper-making contributes only 14.6% of production value, but accounts for 92% of the COD discharged.

In Xinxiang City, even with technology updating to decrease pollutants discharge, the major pollutants discharge amount will still be 3-5 times the environment capacity of the receiving water. Therefore, the water environment protection target must be achieved via industrial restructuring. In this Project, the ability to develop water use and pollution targets jointly, allowed Xinxiang City to develop a specific industrial restructuring plan that could meet water and environmental targets.

(7) Public participation in water resources and environment management

Shortage and deterioration of the quality of water resources have been important factors in restricting economic and social development in the Hai River Basin. While water resources and environment protection departments have adopted strict management measures in the basin for many years, low public participation rate in administrative management has led to lack of incentivesor in "ownership" by the people of water resources development, utilization, and water saving. Under these circumstances it has been difficult to fully enforce all of the policies and administrative measures with the result that regional water resources come under greater stress and the water environment continues to deteriorate. One important mechanism for ensuring public and stakeholder participation in water resources and environment management actions is the introduction of Community Driven Development (CDD) and Water Users Associations (WUAs). Under CDD, WUAs participate in enhancing water rights management and in water-taking permit management. This has been widely demonstrated and promoted in this Project. CDD is created to ensure the ownership of hard-to-choose options by water resource users and by the local community. CDD stimulates individuals and groups to refine their own production service and management innovation; meanwhile, their own capacities are enhanced.

As examples, Guantao County and Cheng'an County, having the most severe water crisis in the basin, were selected in 2005 to participate in CDD activities. Within 3 years, CDD outreach and training included CDD organizational institutions, defining of pilot areas for socio-economic and resource decision-making, developing an environment coordination plan, and establishing project financing and management implementation mechanisms. The result was greatly enhanced democratic participation and ownership of difficult decisions on water use by farmers. In turn, this significantly facilitated the implementation of a regional approach to integrated water resources and environment management which succeeded in relieving the local water resources crisis and guaranteed the sustainable development of that regional society and their economy. On the basis of this experience the CDD experience will be expanded to larger areas in Hai River Basin.

The lessons of CDD in this Project have been that rural communities embrace CDD and are willing and able to make the hard decisions on water allocation and quotas (e.g. ET quotas) once they understand the gravity of their water situation and the options available to them, and realize that they have real choices that are not just imposed from above. Now, they see CDD as a way forward that will assist in sustaining their social and economic situation. However, to make CDD work, the duties, operational procedures, and interactions must be clear to all participants at all level. In this Project, the various duties were:

- County level CDD working group provides support for township and village CDD groups.
- Township CDD working group mainly provides policy guidance, organizational and coordination work for village CDD working group.
- Village CDD working group shoulders more burden, with the major responsibilities such as formulation of village CDD discussion regulations and schemes, drafting of different plans, the implementation, monitoring and coordination of projects and measures in water resources development and utilization, and in management and protection.

Behind this structure are the water resources bureaus at different levels that provide technical support and, in some cases, agricultural bureaus.

Of particular interest in this Project is the broad range of methods and issues that CCD and WUAs could deal with. This included:

- Village involvement using a voting system, to define project establishment financing, implementation, management and monitoring plans, and to ensure the measures are applied.
- Groundwater over-exploration and managing the conflict between water resources supply and demand.

- Unreasonable agricultural planting structure (replacing water intensive crops with less water-intensive crops).
- Water use efficiency – including reducing agricultural and domestic water use.
- More rational use of agro-chemicals to reduce farmers costs and reduce off-field pollution to surface and groundwater.
- Improved storage and disposal of garbage, human waste, and solid waste.
- Choice of options for reducing deep groundwater over-exploitation, for crop restructuring, for ET quota management, and for environmental management and control as shown in the following Boxes.

BOX 7.1: **Groundwater Overdraft Reduction**
Location: Beidonggu Village of Guantao County.
Issue: Excessive use of deep aquifer fresh water with large pumping costs.
Community Decision: Use a mixture of fresh and slightly saline groundwater.
Result: Deep aquifer fresh groundwater pumping was dramatically decreased with reduced agricultural irrigation cost from 38 Yuan/mu/time to 12 Yuan/mu/time yet maintained good economic benefits.

Box 7.2: **Crop Restructuring**
Location: Gaomu Village in Chengan County.
Issue: Existing use of high water use crops such as wheat.
Community Decision: adjust crop structure to less water intensive crops.
Results: Grain and cash crops structure adjusted from 70:30 to 50:50, with 20% reduction in cotton area.

Box 7.3: **ET Management**
Location: Gaomu Village in Chengan County.
Issue: Excessive loss of water from non-beneficial ET.
Community Decision: Allocate an ET quota to individual farm households.
Results: Agricultural field ET is assigned as 711.4 mm, and allowable groundwater pumping is 356,000 m^3. This has reduced groundwater pumping through increased efficiency in irrigation and in control of evaporation from soils.

With the establishment of CDD pilot sites, the following benefits have been documented:

- Change of attitude of farmers with enhanced awareness and ownership of water resources and environmental decisions affecting themselves and their village.
- Autonomous capacity of villagers has been improved through direct involvement in decisions on content, financing, implementation, management and monitoring.
- Enhancement of cohesion and function of WUAs: Water Users Associations

Chapter 7 SUMMARY OF PROJECT OUTOMES AND EXPERIENCE

(WUAs), as one type of grassroots water management organization, is the major force in promoting CDD. In the process of CDD establishment, the WUAs fully play their role and enable farmers to be fully aware of their rights and responsibilities in water resources and environmental management.

- Integrated approach to economic, social and water and environment has been very effective within a CDD approach. As an example, by reducing wheat area and expanding the other crops such as cotton, the cotton planting area in the pilot project area has been increased from 35% to 50%, with an increase in income that is 800 Yuan/mu higher than for planting grains, and with over 50% of irrigation water saved. The quality of life of farmers was improved. This is a win-win situation for economic development and water-consumption reduction.
- The ecological environment has been improved through a more balanced and rational approach to use of agrochemicals, especially for nitrogen fertilizer. In the demonstration area, fertilizer usage has been reduced from 135 tons/year to 110 tons/year – a 18% reduction. Similarly, pesticide use is being replaced by integrated pest management have low residual amounts and rapid decomposition. As a consequence, non-point source pollution is reduced and the ecological environment is improved.

(8) Non-point source (NPS) pollution and countermeasures

Currently in China, with increasing control over point source pollution (industrial and municipal wastewater), the percentage of total pollution contributed by non-point sources, especially from agriculture, is rapidly increasing. In the Project, a major effort was directed to understanding and quantifying the role of agricultural NPS pollution.On the semi-arid North China Plain, due to absence of runoff in most years, fertilizer runoff is minimal except in flood years. However, COD and ammonia from animal raising are an exception and have become major contributors to water pollution. The Project generated a series of recommendations for managing rural, agricultural NPS pollution.

7.6 THE GEF HAI RIVER PROJECT AND THE 12TH "FIVE-YEAR" PLAN

As the GEF Project moved towards its end, the question of how to move the experiences and successes of this project into the planning for the national 12th "Five-

Year" Plan (FYP) (2011-2015), became an important issue. Like the 11th "Five-Year" Plan that placed environmental objectives on a more equal footing with economic objectives, it was anticipated that the 12th "Five-Year" Plan would continue to break new ground for Chinese development by placing even more focus on environmental sustainability. The senior agencies responsible for the Project (MWR and MEP, as well as the World Bank) believed that the lessons learned in this Project have broad relevance across China and that the 12th FYP was the logical vehicle to bring these lessons to a much wider audience.

The drafting of national "Five-Year" plans involves hundreds if not thousands of people across China before the final plan is approved by the State Council. While the plan captures the essential political, economic and social criteria established by the government, the way these are delivered depends very much on how the plan is written and how consensus is developed on the content of the plan. Each lower level of government also creates its own plan to reflect more specific implementation of the directions expressed in the national FYP. Thus, input into the national plan has direct linkages to lower level FYPs that cover the entire country. As some of those experts involved in this GEF Project were also involved in developing the 12th FYP it was important that there be a careful assessment of those outcomes of the Project that made sense to be included either directly or indirectly in the national plan. The following text describes the process taken by the Project on the development of the 12th "Five-Year" Plan. At the time of writing it is too early to identify the impact that these have made as a consequence of being included in the FYP.

The requirements for drafting and implementation of national 12th FYP required (a) the analysis of the feasibility of implementation of GEF Hai River Project outcomes in other key basins in China; (b) determining the consistency between research outcomes and the requirements of the national plan; (c) summarizing those Project outcomes and key technologies worthy of promotion into the 12th FYP. The products of these actions included:

- **Identified Key Technologies:** Guidance was developed for promotion and application of many of the innovations of the Project including: (a) technical issues, such as KM, that support integrated water resources and environment management approach; (b) remote sensing ET technology applications and ET management systems; (c) integrated land-sea approaches whereby actions on land would have

beneficial effects on the adjacent sea area; (d) development and application of integrated river coding systems.
- **Inter-sectoral Co-operation:** This included (a) multi-departmental coordination mechanisms; (b) top-down and bottom-up cooperation mechanisms.
- **GEF Hai River Project Outcomes Promotion Plan:** This activity focused on developing a plan that would promote Project outcomes. The Plan also defined how promoting the successes of the Project should be done (e.g. training, publishing, consultation, etc.), what the contents should be, promotion schedule, experts to be involved, etc..
- **KM System Promotion Consultation Plan:** Development and application of the KM system is the important technology outcome of GEF Hai River project, which provides an effective management tool for basin level, sub-basin level, provincial (municipal) and project county level (city, district level) water resources and environment management. The comprehensive promotion and application of KM system in provinces (municipalities), counties (cities, districts) will be one important task and outcome. Therefore, it was important to draft KM system introduction materials and to provide technology training and consultation services for KM system functions in operational water resources and water environment management.
- **Hold Expert Consultation Conferences on Key Outcomes:** To adopt the technology outcomes generated from GEF Hai River project into basin and regional IWEM planning and management it was necessary to organize briefings, consultations, workshops for relevant departments, basin management institutions, research departments, domestic and foreign experts, including the water resources and water environment management departments and planning drafting departments at different levels of government. In particular, the MWR and MEP have agreed to further promote the outcomes of the GEF Hai River Project.